Little has been written about honor in the social sciences and almost nothing about grace. Yet honor has caused more deaths than the plague and grace is what we all yearn for, whether in the form of favor, luck, pardon, gratuity, or salvation.

This collection of essays develops a line of thought in anthropology which was opened in the 1960s by the editors (and some of the same contributors) in *Honor and shame: the values of a Mediterranean society*. The essays, half of them historical and half contemporary, deal with different aspects of honor and grace, and the strategies and transactions by which they can be obtained. They range from the French royal rituals of the Middle Ages to the murderous feuds and peace-making rites of the Rif; they show how different peoples and periods have faced the problems of power, legitimacy, purity, divinity, and personal destiny. The concluding chapter suggests that anthropology, which ignored honor until a quarter of a century ago, should no longer ignore grace, whose varied connotations provide the basis of religious doctrines as well as the common coinage of the exchange of favors and thanks.

Cambridge Studies in Social and Cultural Anthropology
Editors: Jack Goody, Stephen Gudeman, Michael Herzfeld, Jonathan Parry

76
Honor and grace in anthropology

A list of books in this series will be found at the end of the volume

HONOR AND GRACE IN ANTHROPOLOGY

Edited by J. G. PERISTIANY
and
JULIAN PITT-RIVERS

The right of the University of Cambridge to print and sell all manner of books was granted by Henry VIII in 1534. The University has printed and published continuously since 1584.

CAMBRIDGE UNIVERSITY PRESS
Cambridge
New York Port Chester Melbourne Sydney

Published by the Press Syndicate of the University of Cambridge
The Pitt Building, Trumpington Street, Cambridge CB2 1RP
40 West 20th Street, New York, NY 10011-4211, USA
10 Stamford Road, Oakleigh, Victoria 3166, Australia

First published 1992

Printed in Great Britain at the University Press, Cambridge

British Library cataloguing in publication data

Honor and grace in anthropology.
1. Honour. Mediterranean region. Anthropological perspectives. Study regions: Spain. Andalusia
I. Peristiany, J.C. (John George), 1911–87
II. Pitt-Rivers, Julian
306

Library of Congress cataloguing in publication data

Honor and grace in anthropology / edited by J. G. Peristiany and Julian Pitt-Rivers.
p. cm. – (Cambridge studies in social and cultural anthropology: 76)
Includes index.
ISBN 0 521 39055 9
1. Mediterranean Region – Social conditions. 2. Honor.
I. Peristiany, John G. (John George), 1911–87.
II. Pitt-Rivers, Julian Alfred. III. Series.
HN380.M4H66 1991
303.3'72.091822–dc20 90–1861 CIP

ISBN 0 521 39055 9 hardback

WD

Before this book had gone to press John Peristiany was smitten by a sudden and unsuspected illness to which he succumbed. Let it remain, therefore, as a monument to a man to whom all Mediterranean scholars owe much.

Contents

Illustrations

Notes on contributors

John Peristiany, who died in 1988, was senior lecturer in Social Anthropology, Oxford, 1949–63; United Nations Professor of Social Science, Athens, 1960–78; and one time Ambassador of Cyprus to France. His original fieldwork was on the pastoralists of East Africa; later he concentrated on the anthropology of the Mediterranean. His edited collections include *Honor and Shame: The Values of Mediterranean Society*, *Mediterranean Family Structures* and *Mediterranean Marriage Strategies*.

Julian Pitt-Rivers is Directeur d'Etudes, Ecole Pratique des Hautes Etudes, Paris and was Professor of Anthropology at the London School of Economics and Political Science. His fieldwork has been done in Andalusia, Chiapas (Mexico), the Andes, and southwestern France. He organized the first conference on Mediterranean social anthropology in 1959.

Catherine Lafages studied anthropology at Paris X, Nanterre.

Emmanuel Leroy Ladurie is Professor of History at the Collège de France and Administrateur-Général of the Bibliothèque Nationale.

Pierre Bourdieu is Professor of Sociology at the Collège de France and also founder and director of the review journal, "Actes de la Recherche en Sociologie".

Julio Caro Baroja, historian and anthropologist, is a member of the Royal Academy of Letters and the Royal Academy of History, Madrid.

John Campbell, anthropologist and historian, is a Fellow of St Antony's College, Oxford.

Maria Pia Di Bella is a member of the Centre National de la Recherche Scientifique. She has worked on Sicily and southern Italy.

Raymond Jamous is a member of the Centre National de la Recherche Scientifique who has worked on Morocco and on India.

Sandra Ott is a fellow of St Antony's College, Oxford and is an expert on the Basques.

1

Introduction

J. G. PERISTIANY and JULIAN PITT-RIVERS

This is a collection of essays by authors of different nationality and discipline on a theme that preoccupied us already in the 1960s. It is the continuation, in much the same form, of an earlier collection, but it concentrates upon an aspect of the subject we think we had somewhat neglected before: the relation between honor and the sacred.

The previous book, *Honor and Shame* (Peristiany 1965),[1] was the first attempt to examine the concept of honor in anthropology and it was also, subsidiarily, a preface to the anthropology of the Mediterranean. The book was composed of essays written mostly after, but all as a result of, a conference held at Burg Wartenstein by the Wenner-Gren Foundation in 1959, and as such it was one of two publications, the other being entitled *Mediterranean Countrymen* (Pitt-Rivers 1961). This last, the earlier to appear in print, was concerned mainly with the diversities of the culture of Mediterranean rural communities and with their various ways of adapting themselves to their common problem of belonging to a centralized state and a "Great Tradition," a problem that, with the exception of Robert Redfield, had largely been shirked by anthropologists of previous generations. The second, *Honor and Shame*, sought to define a field in which we had discovered at the conference a certain basic cultural unity.

Looking at honor again in the 1980s enables us to see it in a rather different light, not only because the authors of *Honor and Grace* include new (and younger) scholars, but because those of the first book have found more to say about the subject. These essays were, with two exceptions, written specially for it and were presented, in embryo, in Julian Pitt-Rivers' seminar at the Sorbonne (and their bibliographies have not been revised). Those of Bourdieu and Leroy Ladurie were written at the same time, but independently, and were discussed in the seminar. They have already appeared in the original French (Bourdieu 1986; Leroy Ladurie 1983).

In the years since the publication of *Honor and Shame* anthropology, also, has somewhat changed; it has become much more concerned with symbolism and religion. Hence it appears today that in concentrating upon a social explanation of honor most of us – Campbell was the exception – had been to some extend blind to its intimate connection with the realm of the sacred which was mentioned, then, only as a residual aspect invoked to explain how it was attached to the authority of the State, the Church or the religious leaders. This time, the main orientation is provided by that aspect of honor which faces the spiritual world, the ineffable and the problem of individual and collective destiny.

This change in perspective is due not merely to a change in fashion in anthropology, but to the impossibility of making a complete analysis of honor without examining its ties with ritual and religion. The ties with ritual are obvious, for rites establish consensus as to "how things are" and thus they fix legitimacy. Hence ritual is the guarantor of the social order, conveying honor, not only in the formal distribution of dignities on ceremonial occasions, but also in the sense of making manifest the honorable status of the actors, whether in the rites of salutation, as Firth (1936) made clear, of hospitality, or of passage. The latter, as Bourdieu points out in this volume, are frequently concerned with distinguishing between those who pass the entrance exam to one of the "Grandes Ecoles" and thus receive honor, and those who do not. (Appropriately, an examination is something that, in English, you "pass" or "fail".) But the tie with religion is less evident, for it goes much deeper than the polemics between the nobility and the Church which occupied our attention in the previous book.[2] It is concerned with the notion of grace, both in its religious context and in its daily usage. Hence the title of the book.

It is above all its relation to the ultimate source of the sacred within each individual that brings honor into the religious sphere. It is in this sense that a person's honor is said to be sacred, something more precious to him than even his life, of which it was traditionally viewed as the epitome. "Rather death than dishonor!" was the ideal expression of this sentiment, whether on the battlefield or in the boudoir.

One is perhaps chary of talking loosely about "the sacred" today, for the notion has been overworked since the days of Durkheim, yet we nevertheless require a word to denote the fact that there is a realm of mental behavior where the extraordinary is opposed to the ordinary and where accepted truths are placed beyond question by a conviction, impervious to reasoning because it lies deeper than consciousness; it commands an attachment that springs from the bottom of the self, refusing the logic of everyday conduct. This is where the sentiment of honor arises, and whether or not this realm is claimed by a particular religion as part of its domain and placed under the aegis of its divinity, it clearly borders on the territory of religion and it is in this sense that we can speak of an aspect of honor related to the sacred.

In addition, the paradoxes of honor, seldom recognized by people save when it is a matter of making a fool of someone by playing on the different implications of the word, endow it with a mystery that places it out of bounds to practical reasoning. The Church has always claimed to annex it to doctrinal issues by making it equivalent to "moral conscience" but it clearly is not that in the view of most laymen. Common parlance allies honor to the sacred at large rather than to formal religion, and the rituals that commit it imply as much. The fact that, in English, honor should be expressed in a number of contexts by the notion of grace – to receive grace is to be honored and "disgraced" is simply the synonym of dishonor – carries the same implication, and it is not much different in any romance language. Consequently honor has a kind of congenital relationship with grace which challenges our previous view of it as primordially a matter of social structure.

But we are concerned with more than verbal equivalences. Therefore the fact that on the far side of the Mediterranean we find another concept, not by any means identical with grace, yet of the same order – *baraka* – encourages us to search on a more general level for the logic of that complex of notions that connects status, power, and authority on the one hand with good fortune, legitimacy, and sanctity on the other.[3] For that reason we have also included the Basque concept of *indarra* among the notions examined, for it is synonymous with neither grace nor *baraka*, presenting a rather different conceptual configuration, yet belonging to the same general category of concepts that hold a society together morally.

The sacredness of power is an old theme in political science. If these studies can bring anything new to the subject, it is because they provide concrete examples analyzed within an anthropological framework which combines the deep-seated premises of the culture in question with the social organization of the society.

Casting a momentary glance at the history of the concept of honor, we can see that the word had very little currency in anthropology prior to the 1960s. Mauss mentioned it in passing and even saw the similarity between honor and the *mana* of Polynesia, but otherwise it was rarely invoked in the literature of the discipline, either because honor was too sacred a notion to be credited to peoples who did not belong to Western civilization, because honor was thought to be specifically an element of Western culture not translatable into another language,[4] or because nobody but Mauss had perceived the resemblances between the honor of Maori chiefs or Kwakiutl potlatchers and that of Western men of honor, who had not yet at that time been recognized as a valid object of anthropological study.

In fact the first analyses of honor dating from that epoch opened up new problems: it was evident that honor as a concept was part and parcel of the culture of Mediterranean societies, whether Christian or Islamic, but it was also evident that it was related at the level of behavior to other aspects of those societies, to the values attached to the sexes, the so-called endogamy or the institution of patronage, etc., over and above the simple matter of insult and retaliation which provided the pretexts for validating claims to precedence or for inflicting humiliation by denying them, the common currency of "affairs of honor." In each of these different institutions a different aspect of honor is brought to the fore and the word can equally refer, in English, to a judge, the acceptance of an invitation, female sexual purity, or the crediting of a cheque, so that one might wonder what they have in common at all save the word and why they use the same word to denote such different matters.

These different aspects did not all fit together like the pieces of a jigsaw puzzle composing a whole picture (in what senses are cultures wholes?). Rather they represented not only the variety of social personalities present in a given society but also the varied and conflicting interests of rival groups: lineages, clans, classes, or guilds which gave precedence in their definition of honor to the aspects which favored their social promotion. Warring interpretations of honor contended through their champions to impose themselves upon the popular consensus of each community or kingdom, simply ignoring the contentions of others, for honor is too intimate a sentiment to submit to definition: it must be felt, it cannot be analyzed except by the anthropologist. It is therefore an error to regard honor as a single constant concept rather than a conceptual field within which people find the means to express their self-esteem or their esteem for others.

Moreover our analyses implied a different kind of relationship between concepts, which are a part of culture, and the social reality which could be observed from outside, a relationship which might be seen to depart from the simplistic notion, favored previously by many, that concepts were, so to speak, precipitated by social relations (see Gellner 1973). This was clear from the moment when we perceived a possible social function to be accorded to paradoxes and ambiguities: culture was no longer a set of rules of conduct followed blindly which supported the organization of society, but a structure of conflicting premises within which the struggle for dominance took place. The relationship between culture and society was no longer concordant but dialectic. The achievement of honor was not then simply a refraction or demonstration of the reality of power or precedence, as Thomas Hobbes (*Leviathan*, chapter 10) thought, but also a means of achieving or maintaining them through the control of the definition of honor. The Church and the nobility had disputed this for centuries and not only in Spain: honor owed to God versus honor owed to arms and their courageous use.[5]

And each class possessed its own honor and evaluated itself and others in those terms, down to the statutorily infamous who had none and whose views were not recorded (unless the picaresque novels are to be taken as evidence).

But the different classes were not only credited with different degrees of attachment to honor of one sort or another, they entertained different conceptions of it. (The essay by Caro Baroja deals specifically with this question.) Hence the power of rank versus the power of cash and the convertibility of one into the other was a favorite theme of the social critics of an earlier century, and sometimes events listened to the critics and imposed change. To give but one example, the Puritan revolution in England saw the most spectacular redefinition of honor there has ever been. Rank was discounted, virtue promoted, financial honesty revered, any form of flamboyance frowned on. The theatres were closed. But the Commonwealth did not last; the Restoration returned, if not entirely, to the earlier concept, as criteria of honor, rank, ostentation, dominance, and *joie de vivre* replaced probity, piety, purity, and sobriety, and the ideal of social equality went out of the window. The theatres reopened and for a decade offered almost no joke other than the indignity of deceived husbands whose misfortune had been unmentionable for a generation.

The paradox that honor is at the same time a matter of moral conscience and a sentiment on the one hand, and on the other, a fact of repute and precedence, whether attained by virtue of birth, power, wealth, sanctity, prestige, guile, force, or simony (or, to take another paradox, that those whose honor is greatest feel least obliged to defend it), implied that honor could not merely be reduced and treated as an epiphenomenon of some other factor, but obeyed a logic of its own which could dispel the paradoxes. The ultimate justification of this book is as an attempt to elucidate this logic through the chosen examples of the essays it contains.

A comment on the reception of the previous book may perhaps help to clarify the issues faced in this one, for it was not always understood in the sense which we intended. There was, inevitably, a reaction in defense of a previous and simpler view of the relationship between culture and society; material reductionist arguments were put forward to show that honor was no more than a manifestation of the distribution of power or wealth. It was suggested that the value attached to the sexual purity of women as the basis of their honor could be explained by their role in the process of production or reproduction, or the principle of patriliny, etc. Thus the paradoxes were evacuated, unresolved, by the simple procedure of treating them as subsidiary or fortuitous.

Honor and Shame bore the subtitle "The Values of Mediterranean Society." *Honor and Grace* is no longer concerned specifically with the Mediterranean, for it contains only one essay on the Islamic world and deals with the northern half of medieval France (and England), and also with the Basques, who can hardly be described as Mediterraneans. Yet it is a continuation of the first volume in that it is centered on the same part of the world and concerned with the same theme, of which it attempts to develop an aspect inadequately treated in *Honor and Shame*. But the fact that the subtitle of the book mentioned

"Mediterranean Society" (deliberately avoiding the word "culture") nevertheless led sometimes to the misunderstanding that we were proposing to establish the Mediterranean as a "culture area." This was not the case. This methodological concept did sterling service in organizing the early ethnography of the North American Indians, but its value is much less when applied to a functioning society about which much more is known, for it becomes arbitrary and depends upon the aspect of culture which is given preference in determining the "area." Are the Serbs part of a Mediterranean culture area? In many ways, yes! But from the point of view of kinship, not at all, for until quite recent times their clans were exogamic (see Pitt-Rivers 1983: Introduction; 1986). The Sahara and the Alps set limits to the north and south, because people are very noticeably different on the far side, but where does it stop if you go east? In fact we were as much interested in the differences of culture as in the similarities among the peoples surrounding the Mediterranean.

Our aim in treating the Mediterranean as a whole was epistemological only and we never attempted to define it geographically. Thus when mention is made of "the Mediterranean concept of honor" we recognize that it indicates a tendency to associate masculine honor with female sexual purity only rather vaguely, for there are areas near the Mediterranean where this connection is not made at all, for example certain areas of northern Spain or the Mzab or Ouled Naïl of Algeria, and that in any case there is considerable variation within what is loosely termed so.

Another misunderstanding arose from the conclusion, reached by some readers, that the use of the word "shame" in the title implied that we considered the culture of the Mediterranean was a "shame culture" rather than a "guilt culture," a distinction that had been launched originally by Margaret Mead, but that had convinced us of its utility no more than it had Alfred Kroeber (1900: 612ff.). We might perhaps have done better to entitle the book "Honor and Infamy" rather than *Honor and Shame* in order to guard against this misapprehension, but this would have impoverished the theme since, unlike honor, infamy refers uniquely to reputation and not to sentiments. While admitting all the interest that a study of the psychology of shame may have, especially in relation to the physiology of blushing and other psycho-physiological phenomena, we were content to limit ourselves strictly to the field of social anthropology and confine our investigations to the role of the concepts of honor and shame in ordering society rather than confront the ill-defined problems of the personality of Mediterranean peoples. It appeared to us that the distinction had been lifted from the popular moral philosophy of the anglophone countries and applied in fields where its relevance was dubious, for the sentiment of shame appears to be universal and guilt is simply internalized shame, so that though the degree of internalization of the sanctions varies, no culture can be unequivocally called one or the other. In addition, the subject which tends to be internalized varies: Western Europe from one period to another internalized its sexual shame in different degrees; the Protestant ethic, in addition, internalized financial shame. The Trobrianders feel it on neither of these counts, but many primitive peoples feel guilt with regard to the violation of taboos which Europeans regard as morally neutral. To fear the punishment of the ancestors might or might not be interpreted as feeling guilt. Strathern raises the question, "Why is shame on the skin?" (in Blacking 1977) and concludes that the natives of Mount Hagen, New Guinea, feel both guilt and shame and that the same association between shame and skin is present in Italy, Spain, or Greece: to have *cara dura* in Spanish is to be insensitive to the sanction of shame (it is not far from the notion of a thick skin in English). Yet susceptibility to shame does not entail either the presence or the absence of guilt. The opposition then is a false one as far as cultures are concerned, however useful it may be in the discussion of individual personalities within our own society. But there is a more serious difficulty: one feels guilt in relation to those to whom one is committed. No gypsy ever felt guilt in the face of a *gadjé*, but this does not mean that he is necessarily incapable of such an emotion in relation to his own kinsman, that is to say, towards a person with whom he shares a moral bond and common values. Gypsies, of course, feel no such bond with those who are not of their race. But even to distinguish between guilt and shame as between an abstract moral sentiment and an attitude towards specific persons encounters difficulties: for one can feel guilt in relation to a person and theologians have written of "religious shame."

Finally the attempt to distinguish between guilt and shame cultures faces a last problem: there is no single attitude towards sin and culpability in any one Mediterranean society. In keeping with its definition of honor the Church expected a sentiment of guilt to be aroused by recognition of dishonorable conduct and offered forgiveness through confession, while the nobility tended to defend itself from being put to shame by drawing a sword. Conceptually intermediate between the two, the peasantry strove, rather than for individual supremacy, to maintain the ethic of equality within the community, to avoid being excluded as inferior. As Peristiany put it (in the Introduction to *Honor and Shame*, p. 14), "This is a society without clearly defined spheres of competence in which each male individual constantly asserts his basic equality with all others. The Greek Cypriot Highlander is not antagonistic but agonistic."

Nevertheless we can see, in this example, that the dispute between the two "estates," noble and religious, was not simply a question of the interests of either, but derived from a difference of moral conscience and of values in their approach to the dual nature of the notion of honor: the Church related the sentiment of honor to a conscience free from guilt; true honor was only in the eyes of God; the subjective aspect was the essential, for this lay within the jurisprudence of the Church which controlled only the sanctions deriving from religion. But the nobility, the wielders of military power in origin and subsequently the class entitled by their rank and material possessions to exercise sanctions, had a more worldly conception of honor, for they were concerned to maintain the pecking order (Pitt-Rivers 1977: 3) and defend themselves from humiliation, to avoid shame. Those who exemplified guilt culture were then by implication supporters of the definition of honor which derived it from virtue, those of shame culture took a view somewhat closer to that of Hobbes and interpreted it as above all a matter of precedence. Guilt related to the lack of virtue, shame to the loss of precedence or face. In fact both approaches appear to be necessary, and in a sense complementary, since the function of the concept of honor is precisely, despite the frailty of the logic involved, to equate them and establish thereby the dialectic between (in Pitt-Rivers' words) "the world as it ought to be and the world as it is." But the idea of reifying the distinction between guilt and shame and attaching it specifically to one culture or another removed the possibility of understanding either, since the first essential step in such an endeavor is to examine the terms of their coexistence within a single culture. The Church and the nobility were thus, in this sense, as necessary to the society as a whole as the brahmin and the Kshatriya were to Indian society, even though the relation between them was quite different. This relationship, at the same time one of opposition and interdependence, and their ultimate reunion in the figure of the monarch is illustrated in the fact that the two came together at the apex of the medieval kingdom in the coronation rite in which the king was crowned and inducted simultaneously as knight and as bishop, chief of the temporal order as of the spiritual order, arbiter of honor, his decisions validated by his grace.[6]

This book is, as we have said, about power, success, legitimacy, purity, and sanctity. The opening essay by Catherine Lafages provides a clear, explicit model for such connections. Just as our earlier analysis explained the notion of honor as both a moral and a social matter, and showed how the two aspects were related, so the royal rituals of succession bring the notions of honor and grace together. Without Divine Grace the monarch would have no right to rule and were he not legitimate Divine Grace would surely not descend to defend him – he must be *Red Dei Gratia* or he is a tyrant doomed to a bad end – and the performance of these rites makes the passage of grace patent, hence the importance of respecting every ritual detail and above all, in medieval France, of getting crowned at Reims and anointed with heaven-sent oil. Only Napoleon did not bother to go there for his coronation, for he was crowned not king but emperor, and in any case the whole thing was a muddle, in terms of ritual, from the traditionalist viewpoint: the Pope was there, but he did not perform the rite, for the emperor took the crown out of his hands and placed it on his own head himself, thereby demonstrating in the view of legitimists that he was not a ruler by divine consent but a usurper.[7]

From his own angle this ritual innovation demonstrated that he was a self-made ruler depending not on Divine Grace but on his personal charisma and his soldiers' loyalty. He was not aspiring to be king of France but emperor of the French.

According to the traditional rite, temporal power and spiritual authority are combined, each lending support to the other. The grace of God is necessary for the exercise of legitimate rule and its "gratuity" ensures that the king can do no wrong, since grace implies the right to be arbitrary, free of obligations of a contractual nature; it places him above the law of which he is the defender as it applies to others. The sword, necessary to enable the king to defend the faith, is sanctioned by grace. The ritual represents symbolically the politico-religious theory: the knight, whose privilege (and function) is violence is assured of the legitimacy of his rule to impose peace by the grace which he received as bishop, and since grace is also pardon he receives, in advance as it were, forgiveness for the use he may make of his sword. As Lafages sees it, not only must he be prepared to shed blood in order to defend the divine order upon earth, he sheds it symbolically by the very fact of succeeding to the kingship.

To defeat in a game is to "eliminate," symbolically to kill; replacement is also killing and at the same time incorporating, hence in many African kingdoms the new king is said to have "eaten" his predecessor. Freud has shown how successors feel quite unjustifiable guilt for the death of their parents and the symbolic aspect of rites is often closer to the unconscious than to the conscious. In the symbolism of the coronation the new king is always in this sense a "King of the Wood," murderer of his predecessor, and, qua individual monarch, destined to be destroyed by his successor. Kantorowicz's analysis of the "king's two bodies," the one mortal and individual, the other collective, representing the deathless kingship, showed how this doctrine provided the means of surmounting the contradiction between the mortality of the living monarch and the survival of that other body, the immortal one, existing materially only in effigy in the mortuary ritual.

Lafages shows how the double persona of the king, consecrated as both knight and bishop, with temporal and spiritual power, is able to unite in his person the two aspects of the kingship, the mortal (and individual) and the eternal (and institutional), represented by the two bodies – until the day when death separates them and the individual must be replaced. But since he is also immortal, invulnerable to natural death, he can only be replaced by murder, like the King of the Wood, or like those smothered kings of Africa who cannot be allowed to die a natural death lest the kingdom perish with them. But it is only the mortal king who is slain, the immortal one lives on in the physical person of his successor, a fresh mortal who has incorporated the deceased ruler.

Hence the necessity for that instantaneous transition represented in the proclamation: "The King is dead, long live the King."

It is not necessary to accept literally Frazer's ideas on the murder of the divine king, nor Freud's on the universal wish to destroy one's father, in order to see that these symbolical manipulations maintain the continuity of the kingship over time as a single person. Moreover, Lafages' interpretation of the symbolism looks much less arbitrary when we come to Jamous' study of the *baraka* (which is heavenly peace as well as the assurance of health, success, progeniture, etc.) of the new sultan, religious leader as well as temporal ruler, "Commander of the Faithful" (Waterbury 1970). The new sultan's *baraka* must be literally assured, as well as demonstrated, by the shedding of blood. By virtue of the ancient anthropological principle of the ambivalence of power – "he who can kill can cure and vice versa" – to prove your power to make peace you must first make war. The ritual solution of an insurmountable contradiction between the mortality of the incumbent and the immortality of his function echoes the paradox of honor, which is at the same time a matter of the facts of power and also of moral conscience, of what is and also of what ought to be. They are, as it were, "forced into assimilation," their incompatibility placed in parentheses, just as the facts of power and the religious ideal were exonerated of their contradiction by being encapsulated in the same word, "honor," rendered thus indistinguishable, for the greater glory of God and the greater confusion of poor mortal scholars.

Louis XIV, whom we encounter in the next essay, Leroy Ladurie's, is a very different kind of monarch. Though he was crowned at Reims and anointed with the oil of the holy phial, his sacrality has a different feel to it. He is sacred by his birth rather than by his coronation, for in the meantime certain things have changed: the State has come into existence and the rule of succession has become more strict than it was in the days of those medieval kings who divided up their possessions at their death between their sons and, corresponding to this change, the doctrine of the Divine Right of Kings has been formulated – and even before Louis' birth it has already cost his "cousin" Charles I, across the Channel, his head. The Fronde and the wars of religion have all but undone the kingdom of France when Louis is still a child and, if the menace remains of the return to anarchy if ever his legitimacy were placed in doubt, the State's authority has finished by imposing itself; it is centered now upon the person of the king. This is strikingly visible in the elaboration of ritual surrounding his physical person: dressing it or undressing it, or rather redressing it, for bed, ceremonies almost as complicated as a coronation, and designated by the words which describe the rising and setting of the sun: *lever* and *coucher*. It also lies at the basis of his concern for the advancement of his bastards, the fruit of his loins, which so deeply offended Saint-Simon. It is not that royal bastards were without social and political importance in earlier times but Louis XIV's are institutionalized as the element of his family that is closest to him by blood and therefore should be ranked highest in court etiquette after the legitimate of the same degree of affinity. He is sacred in himself, not just by the Grace of God with which he had been anointed. Within a pecking order, controled meticulously by himself, he feels free to advance his illegitimate children, to the fury of the "Little Duke" who vents his spleen (of which he has a generous endowment) in his immortal, if unpublishable, memoirs. The nobility have been shorn of their regional roots and their local support, and Richelieu has knocked the fortifications off their castles; they now depend upon royal favor. Much the same can be said of the Church in whose affairs the royal mistress plays an overbearing part.

Power is still in need of the sacred in order to assure its legitimacy, but the sanctity of the royal person is enough to sanction all the innovations of the reign. The pecking order, manifest in the chapel at Versailles, is fought out in terms of privilege and ritual detail. Much of the ancient ritual remains, but its significance has changed: the holy cloth is still held by its four corners to represent the four corners of the world, but the centre of interest has shifted: to focus upon which duke holds which corner.

Three centuries later the monarchy has gone, but not the need to legitimize the rulers. Yet the way this is done has changed: there is no more holy oil and no rights attach to birth, though birth provides the necessary "background" to pass the rite of passage by which the new rulers are selected for the roles of power, are "instituted" in Bourdieu's term. Nobody speaks of honor any more except in public speeches on ritually loaded occasions, yet precedence and virtue still compete as criteria for recognized honor: the Légion d'honneur leans upon a mixture of public renown and private favor. The consecration of power that Bourdieu explains so pertinently still operates even if the idiom has been secularized and grace has been stripped of its religious associations.

Thus the book starts with three essays on the relationship of honor and grace and their role in the legitimization of authority, illustrated by three very different periods of French history, and a fourth historical essay follows.

With Julio Caro Baroja we move from France to Spain and back to the age of Louis XIV, not to deal with monarchy any more, but with the different social "estates" and the expectations entertained by churchmen as to the type of sin each was liable to commit. To these sins correspond, as one would imagine, a particular position in society and a concomitant view of their honor, most visibly of all in the aristocratic proclivity to violence. The power to pardon (*gracier* in French) remains in clerical hands and redemption is obtained from God through the intermediary of the priest. The diversity of the varieties of honor as of sins composes the casuists' view of the social structure. And there are nuances of difference between the secular and regular clergy and even between different orders.

The subsequent studies are ethnographical descriptions of the notions of honor and of grace in different parts of the Mediterranean region. All of them deal with one or other aspect of the sacred side of honor. This is not always expressed by the word "grace" or its derivatives, for situations of a similar nature are not always formulated in the same words. In fact we find these notions present in the assumptions that underlie conduct and judgments rather than in words, without which a great deal can be said about honor, starting with the gestures of contempt or desecration (French: "le bras d'honneur"; Italian: "fare li corna"; Spanish: "corte de mangas"). Equally the idea of grace does not always require to be expressed by the word itself.

Peristiany discusses the politics of consensus in a Cypriot highland village where the importance of honor looms large both in terms of the standing of individuals and the sanctions which apply to behavior. In this tightly knit and introverted community, their collective renown also matters a great deal to them and they find it necessary to repair their reputation as the toughest businessmen on the island by occasional grandiose gestures of generosity in favor of the Church, somewhat in the same spirit as the Spanish merchant of the seventeenth century, cited by Caro, who, after a lifetime of sharp-practice, endows a hospital. After all, you have to get the money first before you can be generous with it.

The *sophron* is the arbiter of honor both in his judgments of individual conduct and in his defense of the collective reputation. He must be the epitome of honor himself for people to accord him this position of supreme influence, yet he no longer struggles to advance his personal interests, but acts with extreme caution in regard to the honor of others in order to establish the consensus of the community. For honor varies completely not only according to the sex of the person, but also according to his age. The young man must compete for prestige, impose himself on others, win renown, but with age a different kind of honor is required: the precedence that he established in his youth is the initial basis of his honor, but to make him the spokesman of the community, the epitome of wisdom, his conduct must change with the years. The ideal of the patriarch is to be without possessions, having given all away to his children, to the Church, to the poor. The behavior required by honor is reversed. As in sacrifice, dispossession converts "having" into "being"; the struggle for achievement is over, material possessions are traded for grace.

This essay is followed by Greeks of a very different stamp, the brigands, called "klephts," of the epoch preceding national independence. Outlaws of the mountainous regions, they have their honor as such which obliges them to commit all manner of atrocities on the unfortunate persons who fall into their clutches. They resist the forces of the Turkish government whose power peters out when it leaves the southern plain and whose legitimacy they refuse to admit. Their heroic qualities, supernatural strength, and stamina which they share with the heroes of the *Iliad* (who are all descended from a god) are attributed in their case, also, to a sacred source. They are rebels, but not revolutionaries and they can be reintegrated when the regional governor changes. The "armatoles" who combat them in the name of the government, recruited originally from among their ranks, are in fact mediators who can become klephts again when this happens. They occupy the margins of ordered society, but since the government is Turkish and therefore Muslim, it cannot in any case be legitimate in Greek eyes. The case reminds us therefore that legitimate rule has not only its geographical limits where the power to enforce itself is lacking, but its cultural limits also.

Di Bella's analysis of honor in Sicily goes back to its roots in the notions of name and blood. These correspond to masculine or feminine, positive or negative; honor in that name is, to begin with, social and inherited legally, and is accreted by the acquisition of prestige and precedence by the members of the family that bears that name, or it is depleted by their failure to impose themselves and maintain their precedence. Blood, on the other hand, is women's business and it is a negative quality, since, like purity (and this is what it means: purity of blood) it cannot be gained in any way, but can be lost by positive action: illicit sexual activity. It is conserved by avoiding defilement or any conduct that implies frivolity. Yet this does not mean that only men are concerned with name and only women with blood – rather, the men are the most concerned with the purity of their blood in this sense and the women of the family connive to hide from them the shortcomings of a member of their sex. On the other hand it is the women who demand that their menfolk take positive action in vengeance, to defend their name. The two concepts are related by the fact that the seduction of a woman adds to the renown of the seducer and tarnishes the blood of those who were charged with her protection. In the agonistic struggle to achieve honor through imposing oneself aggressively, all cannot succeed.

There is, however, another road to prestige, the road of grace, through a miracle accorded by a saint. Miracles provide a quite separate possibility of acquiring renown, but renown of a different kind. It is not through triumph on the plane of day-to-day relations that one's superiority is established but by a preferential relationship with a saint whose intercession obtains divine grace. The recipients of miracles come from the other extremity of society: these make no claim to honor as precedence, nor are they even necessarily persons noted for piety and good works, but they must have devotion to their spiritual benefactor with whom all sorts of pseudo-contractual arrangements can be made in the form of vows and "promises" in the hope of receiving a miracle. But the ways of God are mysterious and grace does not respect contracts. Nevertheless, once a miracle has been accorded, expressions of gratitude in the form of ex-votos are put up in the Church, to give thanks and also incidentally to give full notice to the community of the honor received from on high. Those who are excluded from the struggle for achievement may well succeed better in this passive manner than those who strive. Sanctity is attained at the cost of abjuring violence as the means to achievement.

In the Rif we are back in marginal mountains once again with Jamous' study of honor and *baraka*. Here in "lands of insolence" where the rule of the law of the state does not penetrate we see perhaps more clearly, from the margin, what the center is really like, though this time the religion is the same as that of the rulers: the Sultan rules spiritually even where he does not manage to collect taxes. But we find a similar relationship between violence and peace. Peace is always above violence because the renunciation of violence implies spiritual superiority; the holy men, *shorfa* (sing. *shereef*), men of peace, supreme epitome of *baraka*, are above those who enter into the exchange of violence, because *baraka* is above violence. It is they who put an end to feuds. Yet since *baraka* encompasses the whole of life, the profane as well as the sacred, it is also the source of worldly triumph, of success, even in war. The supra-agonistic condition of the *shorfa* entitles them not only to make peace, but by the same token to represent the whole community. *Baraka* is like grace in that it is a divine quality, not answerable to the mundane logic of social relations utilised by men in their struggle for supremacy, but to the will of God.

Ott describes a concept from a very different culture, Basque *indarra*, which she nevertheless compares with grace and above all with *baraka*. It has little theological underpinning, for theology is rarely developed in the vernacular; it is above all lived, not elaborated intellectually, and unlike grace its prime sense is health and vigor.[8] Basque society is not agonistic in the way Sicily, Cyprus, or the Rif are. Basques love to compete in games where the winner gains only prestige; they are ludic rather than agonistic. And the relationship between the sexes is symbolically rather different from that which is commonly found in the Mediterranean. Yet like grace and *baraka*, *indarra* is the source of personal fate, of legitimacy, of health, but also of social order (the house has *indarra* and the house is the fundamental unit of Basque society), of the legal and the divine order, and, unlike grace, of the natural order also (the wind has *indarra* and *indarra* makes the plants grow). It represents things as they are if they are as they ought to be.

In a final postscript Pitt-Rivers examines the notion of grace as an anthropological concept and draws conclusions as to its utility in anthropology and its theoretical implications.

It should be stressed that although these essays had a common point of departure they are not all founded on identical assumptions and, rather than representing stages in a single argument, they are, as was suggested, "materials for reflection." Yet they are all pieces of evidence, examples of how honor and grace come together in a given situation and they all contribute to the concluding chapter where the question of their relationship is posed in more abstract terms.

To sum up, the relation between honor and grace varies from one chapter to another. Thus in the royal rituals of the Middle Ages the complementarity between the two is shown: the king is endowed with both to become the fount of honor as of grace. The supreme power requires to be not only effective through the sword but legitimized by the miter, symbol of spiritual authority.

In the second chapter, Louis XIV's obsession with the ritual aspect of power is displayed. The ritual expresses the divine sanction, guarantor of his legitimacy, and makes clear the order of precedence.

The legitimization of power through the Grandes Ecoles provides a paradigm from which God is missing; a very different ideology consecrates that of the rulers of modern France, yet the necessity for consecration remains. Bourdieu's contribution to *Honor and Shame* analyzed the mechanics of the struggle for precedence in the local communities of Algeria. Here, at the much wider level of the French State, the struggle for precedence takes place in the form of an entrance examination: the reputation that is thus established assures access to power and in a sense sanctifies the status of the successful candidate.

In seventeenth-century Spain the variety of the concepts of honor and how they are attached to the different "estates" are revealed. Here is a society in which honor is statutorily determined as to its very nature, and it is not the same for everybody – a conclusion we reached in *Honor and Shame*.

In the Cypriot example, on the other hand, honor is not the same all the time even for the same person; old man's honor becomes grace and requires quite different conduct and fulfils a different function to that of the young "pallikari."

In the study of the heroes of the *Iliad* and the Greek brigands we return to the problem of honor in relation to political power and legitimacy. The klephts have a touchy and bloody approach to honor, but they are not without piety of a certain sort and they respect the rites of the Church. Thus they have a kind of cultural legitimacy because they are Christian, not Muslim, and true Greeks, not servants of the Turkish power.

Honor and grace in Sicily are presented as alternative roads to renown and they imply very different conduct to achieve this. But the difference is no longer a matter of age, as in the case of the *sophron*, but of class.

Baraka in the Rif enables us to compare this Islamic concept with European grace. Basque *indarra* shows how honor and grace can be related in a different way: sometimes *indarra* would seem to correspond to honor, sometimes to grace. In fact it seems, like Polynesian *mana*, to combine elements of both.

This extraneous example shows how difficult it is to give a concise definition of either. Honor is not something concrete or even constant, yet neither is it just "air" as Falstaff thought, for it has caused more deaths than the plague. Its very ambiguities make it necessary to present different ethnographical (or historical) studies before we start to think about it, rather than attempt to impose a concept that has been derived from another field or from one's personal experience. And grace is even more difficult to define concisely! Therefore we do not start with clean unequivocal definitions of both and follow them throughout. On the contrary it is only at the end, in the postscript (which might be read immediately after the introduction by those especially interested in the theoretical issues raised), that the attempt is made to draw the different paradigms together and find a general principle, delimiting a class of concept to which both honor and grace belong. The debate remains open.

Notes

1 The subject has been further developed in the various international conferences subsequently organized by Peristiany.
2 See pp. 24, 32, 43–7; see also Pitt-Rivers 1969.
3 Cf. de Heusch 1962. This valuable collection of essays is mentioned only as an example of a previous attempt to come to grips with a similar problem. It is obviously impossible to name all that has bearing on the topic and we have limited bibliographical references to those immediately relevant to what is said in the text.
4 Cf. Pitt-Rivers 1977: viii–xi. The same problem was raised in relation to the concept of caste by Louis Dumont (1966) in the final pages of *Homo Hierarchicus* and by those who disagreed with him either because they maintained it was a purely cultural notion, specific to Hindu religion, or because they stood out for a general sociological definition equating it with any form of severe social inequality such as is found in the United States or Latin America.
5 This disagreement regarding honor was not limited to the Catholic Church. The Puritans made the same distinction, as Max Weber (1968) made clear when he quoted Cromwell's "Men of Conscience versus Men of Honor."
6 The monarch was not only *Rex Dei Gratia*, charged with the responsibility to do the will of God, but literally possessing in his person the grace to cure the "King's Evil" through personal contact, as we shall find (see chapter 11) the humblest old women curers doing in the Andalusian countryside. It is the essence of the mysteries of grace that they do not respect the social order. This is made clear in Maria Pia Di Bella's essay on Sicily (chapter 8).
7 David's picture of the occasion is in the Louvre, though it is not to be trusted in every detail: Madame Bonaparte (*mère*) was put in only by the painter.
8 Nevertheless, among the peasants of northern Spain studied by Catedra (1981) to "lose grace" is to lose health and vigor and this loss explains decrepitude and approaching death. Its loss also accounts for suicide.

References

Blacking, John (ed.) 1977. *The Anthropology of the Body*, London

Bourdieu, Pierre 1986. Les Rites comme actes d'institution, in Pierre Centlivres and Jacques Hainard (eds.), *Les Rites de passage aujourd'hui*, Actes du colloque de Neuchâtel, Lausanne

Catedra, María 1981. Death as a cultural process, Ph.D. thesis, Penn University
Firth, Raymond 1973. Bodily symbols of greeting and parting, in *Symbols Public and Private*, London
Gellner, Ernest 1973. Concepts and society, in *Cause and Meaning in the Social Sciences*, London
de Heusch, Luc (ed.) 1962. *Le Pouvoir et le sacré: pour une dialectique du pouvoir*, Brussels
Kroeber, Alfred 1948. *Anthropology*
Leroy Ladurie, Emmanuel 1983. Auprès du roi, la cour, *Annales*, 38 (1)
Peristiany, J. G. (ed.) 1965. *Honor and Shame: The Values of Mediterranean Society*, London and Chicago
Pitt-Rivers, Julian 1961. *Mediterranean Countrymen*, Paris and The Hague
Pitt-Rivers, Julian 1969. Honor, in *Encyclopedia of Social Sciences*
Pitt-Rivers, Julian 1977. *The Fate of Shechem or the Politics of Sex*, Cambridge
Pitt-Rivers, Julian 1983. Introduction to *L'Anthropologie de l'honneur*, Paris
Pitt-Rivers, Julian 1986. Introduction to B. Kayser (ed.), *Les Sociétés rurales de la Méditerranée: recueil de textes anglo-américains*, Aix-en-Provence
Waterbury, John. *Commander of the Faithful*, Columbia, New York
Weber, Max 1968. *On Charisma and Institution Building*, Chicago

2

Royalty and ritual in the Middle Ages: coronation and funerary rites in France

CATHERINE LAFAGES
(translated by Roger Just)

We shall attempt in this essay to determine from the accounts of medieval French (and sometimes English) coronation and funerary rites the logic of sovereignty, and to describe and analyze the manner in which the rituals function both in themselves and in connection with other closely related aspects of royalty.

They form, in fact, a single system, the transition of the monarchy from one reign to the next, though it is convenient to deal with them in the sequence of the individual reign, placing the coronation before the funeral rather than in the order of events which constitutes the transition.

The coronation

Chronologically, the first coronation was that of the Frankish King Clovis in A.D. 496. It was at the same time a conversion and a baptism. From then on this religious connotation remained fundamental to the character of the monarchy. The Holy Ampulla, which was used for the first time on this occasion, remained the essential instrument for unction.

The coronation takes place in eight stages according to Hocart (1969: 98), though I have slightly modified them in my presentation.

1 the preparation of the church and of the king,
2 the combat with, and victory over, impurity,
3 the request made to the king, his reply, his oath,
4 sacred vesture,
5 unction,
6 communion,
7 investiture with the regalia,
8 the acknowledgment: procession and celebration

Preparation of the church and of the King

Two rites of purification are involved: first, everything which is impure must be expelled from the church, and the church must remain empty throughout the night preceding the rite of enthronement. It is guarded by men-at-arms detailed to prevent anyone entering the church before the appointed hour.[1]

Next the king himself prepared by certain operations to "enter into the estate of royal religion." To prepare himself for communion he fasts for three days before the coronation and does penance; that is to say he makes confession in the church the evening before the ceremony.[2]

In another account we read that on returning to the palace the King retires. On the Sunday, once Matins have been sung, the canons line up in procession headed by two spiritual peers (i.e., bishops) whose task it is to search out the King from the neighboring palace where he reposes on a ceremonial bed. The Precentor knocks on the door three times with his staff in the manner of the sub-deacon at the church door on Palm Sunday.

> "What do you want?" says the Grand Chamberlain.
> "We want the King."
> "The King sleeps."
> [repeated]
> The third time they say:
> "We want King Charles the Seventh, whom God has given us as King." Then the door opens and, raised up by two spiritual peers, Charles descends from his bed and is led to the cathedral where the archbishop awaits him for the coronation.
>
> (Baudrillart 1929: 14)

We shall encounter a similar request and reply again, in the context of the funeral.

The combat with, and victory over, impurity

Once this penance and retreat are over, the king is ready to receive the arms which he will need in order to do battle symbolically with impurity, and to defend, not symbolically but in reality, the rights of the Church and of his people. The symbol of the Church's fate entrusted to the King is the royal sword which will bring him a victory analogous to Christ's victory over the enemies of the Church.

The prayer preceding unction, taken from *Le Livre du sacre des rois*, testifies to the combat: "Armed with the helm of Thy protection, and ever protected by the invincible shield, aided by the arms of heaven, shall he [the King] be able to win an enviable victory over his enemies and strike down the infidels, brandishing terror like a sword." In this prayer, the Church is the fighting force, the enemies are the enemies of religion, and the weapon of combat is the sword, which traces a sort of circuit circumscribing within the space of the ceremony the extent of secular power.

After the oath, the sword is placed on the altar in its scabbard. Once the King is undressed and standing in his silk tunic and shirt ready for unction, the archbishop buckles the sheathed sword on him. Then he removes the scabbard, places it on the altar, and puts the naked sword into the hand of the King, who then offers it on the altar. During the blessing the King hands over the sword to the Senechal (or High Constable) who precedes the King holding it point foremost (or upright in certain representations) on his return to the palace. The sword, placed in the King's hands, removed, and then returned to him, consecrates the King as a knight, and the same significance attaches to the presentation of his ring. The prayer pronounced on these two occasions recalls the prayer preceding unction, but on a more spiritual level: "Armed with the protection of the Holy Trinity, may he, as an invincible soldier, continually overcome the armies of the Devil and prepare his soul and his body for true felicity."

This entire section of the ceremony justifies the use of force to conquer power. What sort of power remains to be specified: power over souls. The battle against impurity represents the spiritual aspect of this force, while the struggle against the infidels is its concrete manifestation.

The request made to the King, his reply, his oath

The Archbishop asks of the King: "That he promise and affirm by his oath to maintain and have maintained the rectitude of the bishops and of the churches, as the King must in his kingdom, and other things . . . are put to him to be promised and sworn . . . that is to expel the heretics from his kingdom" (*Le Livre du sacre des rois*).

This oath occurs before consecration, and the King swears it with covered head as a sign of submission. Monsignor Baudrillart (1929) gives his interpretation of the King's oath in the following terms: "The King offers his power to defend the interests of the Church. At the same time he submits to it, and his coercive force is linked to his defense of the Christian people who are under his protection."

This view implies a hierarchical order between Church, King, and people, an order to be found throughout the various levels of feudal society. Thus the Church, in terms of ideological principle, encompasses the King and his people, while the King, as part of his social function, rules over both Church and people.

Similarly, the King of England swears his oath to the constitution in the following terms: "To the Christian people in the name of Christ [I promise] to preserve by our law the Church of God and all Christian people . . . for all time. Secondly, I shall forbid all extortion and all unjust dealings in all ranks. Thirdly, that in all judgements, I shall observe justice and mercy that the indulgent and merciful God grant us His mercy" (Hocart 1969: 95). These conditions were respected during the earlier period of the monarchy, but from the Tudors and under the Stuarts the promise was altered to the point of negating the conditional nature of the King's power as contingent on his protection of the people. The same tendency can be observed in the evolution of the French monarchy with the development of the doctrine of the Divine right of Kings in the sixteenth century.

Sacred vesture

At the conclusion of his oath the King dons his ceremonial vestments which consist at first of three white robes, one on top of the other, later to be replaced by more sumptuous garments of purple or violet. The three important items are the cloak, the tunic, and the dalmatic, which correspond to the attributes of priesthood: priest, deacon and sub-deacon. The tunic is white and bordered with purple like the Roman Emperor's.

It is worth stressing the Archbishop's words to the Dean of Westminster in the English ceremony recounted by Hocart when he places the cloak over the King's shoulders: "Receive this imperial pall which is formed with four corners, that you know the four corners of the world are subject to the Power and the Empire of God, and that no man can reign happily on this earth who holds not his authority from heaven." The "Power and the Empire of God" are associated with the earthly empire, the universe: the "four corners of the world."

The second important point in connection with the sacred vesture is the connection between the three superimposed robes and the three priestly vestments. "In the *ordines* [ritual regulations] at Reims from the thirteenth century onwards the following words are found in connection with the tunic which the kings assume after unction: the tunic must 'be made in the fashion of the tunic in which sub-deacons are dressed for mass'" (Bloch 1961: 202).

The King's cloak is the equivalent of the priest's cope, the dalmatic represents the one worn by deacons, and the tunic is homologous with the dress of sub-deacons. Thus the King is attired with the semblance of the three ranks of the priesthood.

Charles V was garbed in the following manner on the day of his coronation:

> Breeches of violet silk embroidered and worked with fleurs-de-liys in gold. And the tunic [is] of this colour and this same work cut in the fashion of the tunic worn by sub-deacons when they are dressed for mass. And with this the surcoat which must be altogether of the same colour and the same work. And this [is] made very like unto a silk cope without the hood. (*Le Livre du sacre des rois*)

Next the King undresses to receive unction, and he remains "in tunic and shirt," which are left open at the places where he will be anointed.

Note should be taken of the colors associated with these elements in the coronation. The King is first dressed in white, the color of initiation, to receive unction. Then the red cloak, the symbol of sacrifice (the color of rubies) is overlaid, then purple, the episcopal color *par excellence*. The King's successive transformations correspond to stages in his sacralization: the changing of vestments involves a change of state.

Unction

In the account given of the coronation of Charles V, unction commences immediately after the King has relinquished his sword into the hands of the Senechal.

> These things having been accomplished, and the chrism placed on the altar in a consecrated paten, the Archbishop must place the Holy Ampulla on the altar and extract with a needle of gold a small amount of the oil sent from heaven and mix it with great care with the chrism which is set there to anoint the King, which King is resplendent before all other kings of the world in this glorious privilege: that he alone be anointed with oil sent from heaven.

Unction is considered as a sacrament which confers on the King the grace of the Holy Spirit, as a grace given through the intermediary of the oil of the Holy Ampulla mixed in the chrism (though, in fact, as discussed below, Rome was somewhat recalcitrant about recognizing unction as a sacrament).

At this stage let us review the legendary origin of the Holy Ampulla in order to appreciate the full weight of this privilege. The miraculous origin of the Holy Ampulla is related in the 1923 two-volume edition of the Larousse Dictionary.

> It is said that on the coronation day [of Clovis] the priest who was bringing the Holy oils was prevented from arriving in time by the crowd. A dove, which descended from heaven, then brought to Saint Rémi an ampoule filled with a miraculous chrism. The ampoule was kept at Reims, and from the thirteenth century it was commonly believed that although at each coronation a few drops were drawn from it, the level of the liquid never lowered.

In 1793, Rühl, a member of the National Convention, smashed the Ampulla with a hammer on the public square of Reims. Clovis was the first to make use of this miraculous gift. In fact he was not anointed as a King, but baptized according to what was to be called the "Anglican" rite. Legend has turned Clovis' baptism into the first coronation. In fact it was Pepin the Short who inaugurated coronation in A.D. 781 to justify an usurpation of power.

Where was unction received? In 1429 Charles VII received unction on:

> the crown of his head (as in confirmation)
> the stomach
> between the shoulders
> on the right and left shoulder

under both arms and in the crook of both elbows
on the palms of both right and left hands (as in the ordination of a priest).

In the coronation of Charles V the unction on the stomach was made instead on the chest and there was no unction on the palms of the hands. On the other hand, the prayers and formulae accompanying unction all point in the same direction; they are indicative of the privilege accorded the King through the intermediary of the sacrament: "O Christ, consecrate this King for government just as You have consecrated priests, Kings, prophets and martyrs" (Baudrillart 1929: 12). Similarly, when the King's hands are anointed, the prayer is: "Let these hands be anointed with the Holy Oil with which the Kings and prophets have been anointed, and in the same way as Samuel consecrated King David." The medals and representations of the King struck especially for the occasion of the coronation bear the following inscription: "Rex Christianissimus celesto oleo unctus" (Most Christian King anointed with the oil of heaven). Finally those who surround the King sing a hymn during the unction entitled "Innuxerunt regem Salomonem" (They have anointed King Solomon). After the ceremony the Chamberlain readjusts the King's tunic and surcoat, holding the opening with his right hand and the back of it with his left hand.

When the King is reseated, the Archbishop proceeds with the unction of the Queen. She stands on the left-hand side of the chancel, while the King is on the right. She is anointed on the head and chest "not with the King's unction sent from heaven, but simply with holy oil" (*Le Livre du sacre des rois*). The difference in treatment between the King and the Queen can be set out as follows:

KING	QUEEN
unction on entire body: head chest, shoulders, arms and hands;	unction on head and chest only;
with balm from the Holy Ampulla;	with holy oil;
on the right hand side of the chancel	on the left hand side of the chancel.

The parallelism between royal sacrifice and the sacrifice of God is a compelling one. This King is first stripped, bared, disarmed, and placed in an attitude of total humility before both the ceremony's officiant and its participants; then he gains new strength through communion and has the stamp of the Holy Spirit put on him, becomes the very embodiment of the sacred, and, undergoing the sacramental stages reminiscent of Christ, is reunited thereby with the divinity.

During unction, too, there is a recapitulation of a number of rites and of the symbolic significance which goes with them. Unction on the forehead recalls the ordinary unction which the bishop gives the newly confirmed; unction on the arms and palms of the hand becomes one with that given to bishops and priests. But also and most importantly, baptismal unction effaces sin and restores to the newly baptized his original purity. This unction with the miraculous balm of Saint Rémi entitles the King to regain his state of total purity. Equally, it allows him to be completely relieved of the burden of guilt with which he is weighed down. The King is cleansed of his sins: even the stigma of murder is effaced by unction.

Communion

Communion is always accompanied by offering which symbolizes the gift of riches which the King makes to the Archbishop and, through him, to the Church and the people. During the offertory, in the same way as the elected bishop between his two sponsors, the King himself comes to present the traditional offering to the Archbishop, kneeling before him and kissing his ring: wine in a silver vessel, a gilded loaf, a silvered loaf, and "as much gold as he chooses to give" – for the King fixes a certain sum and coins are especially engraved for the day, as coronation medals (Baudrillart 1929: 12).

Charles V's coronation was not dissimilar in this respect. After he had retired into a pavilion situated slightly behind the altar in order to make his confession, the King took communion in both kinds. He received from the Archbishop's hand the host and the consecrated wine contained in the chalice of Saint Rémi. The Queen, too, took communion in both kinds; but neither of them did so until the Archbishop had himself partaken of the sacraments.

Investiture with the regalia

The King, anointed with holy oil, receives his investiture; at this stage he is put in possession of the attributes of his power, the regalia.

The Archbishop takes the crown and places it on the King's head. The spiritual and temporal peers "hold up" the crown, and then conduct the King to his throne. The Queen is crowned in the same way, except that in her case only the barons support the crown over her head.

The scepter shines in the King's right hand; in his other hand he holds the hand of justice (*virga virtutis*) called "the rod." The same attributes, but smaller in size, are conferred on the Queen. A trinity is thus formed: head/crown in the center and above; scepter on the right and rod on the left.

After communion the Archbishop removes the crowns from the royal couple and replaces them with two smaller ones which they will wear until they return to the palace. Further attributes are then given to the King, which constitute part of the arms of a knight.

First are the spurs. These are of gold, and decorated with red stones (garnets), fleurs-de-lis, and buckles in the shape of lions' heads. Supposedly they belonged to Charlemagne. Like the sword, they will be placed on the

King, then removed. Traditionally it is one of the King's peers who fits them on him before unction, and the Archbishop who removes them from him after the ceremony. "First the Grand Chamberlain of France will there beshoe the King with the afore-mentioned shoes which the Abbé of Saint-Denis will place on him. And after that the Duke of Burgundy will place at his feet the spurs which the Abbé of Saint-Denis will fit on him, and which will now be removed" (*Le Livre du sacre des rois*). The sword and the spurs, both integral parts of the armament of a King's knight, undergo the same treatment. They are placed on the King by a secular peer and then removed by the Archbishop. As part of a knight's apparel, they are profane, and they too must pass through the hands of the Archbishop to be consecrated.

Next come the gauntlets, blessed by the Archbishop, and the royal ring, also blessed, and put on the fourth finger, that is the "ring finger " – the marriage finger *par excellence*. With regard to the analogy between marriage and coronation,[3] let us note the following passage taken from the description of the coronation of Anne de Bretagne in 1504 at Saint-Denis: "The Cardinal Legate, Georges d'Amboise, placed on the head of Queen Anne the crown of Saint Louis, and handed over to her the scepter, the hand of justice [the rod], *and the ring intended for wedding the kingdom*" (Pinoteau 1972: 308 n. 13).

This crowning, even though the Queen was not anointed, is referred to as a coronation. The ring is the symbolic mark of the King's marriage with his kingdom, in the same way as within the Church the ring serves to symbolize a priest's marriage with God. Similarly the marriage in France in A.D. 856 of Judith with Ethelwolf, King of England, was at the same time a coronation (Hocart 1969: 103).

As well as the ring and the gauntlets, a gold-covered prayer book containing the coronation prayers and a clasp used to fasten the royal cloak feature amongst the regalia. The scepter and the rod are also attributed to Charlemagne: he is represented on the massive gold scepter as seated, orb in hand, and, notably, surrounded by two lions and two eagles. Also to be noted is the presence of a pink ruby called an Alexandrian ruby. The hand of justice takes the form of a golden baton surmounted by an ivory hand wearing on the fourth finger a ring set with a sapphire. Three wreaths in the form of foliage encircle it. It too is part of Charlemagne's legacy. Though the rod is borne in the King's left hand, it represented a right hand during the Middle Ages. Later, it tends to be a left hand (see catalogue of the exhibition, *Les Regalia*).

Let us now return to the crown and the sword.

The crown used for coronations at Reims is called Charlemagne's crown. It was used from Philippe-Auguste to Louis XVI. The ancient French crown is one of the heaviest, and weighs four kilos. Hervé Pinoteau stresses its weight and the distribution of its stones in his study of it.[4]

The crown used to crown the kings "has a very rich tiara garnished with

Figure 2.1 The hand of Justice, from the Honors of Charlemagne

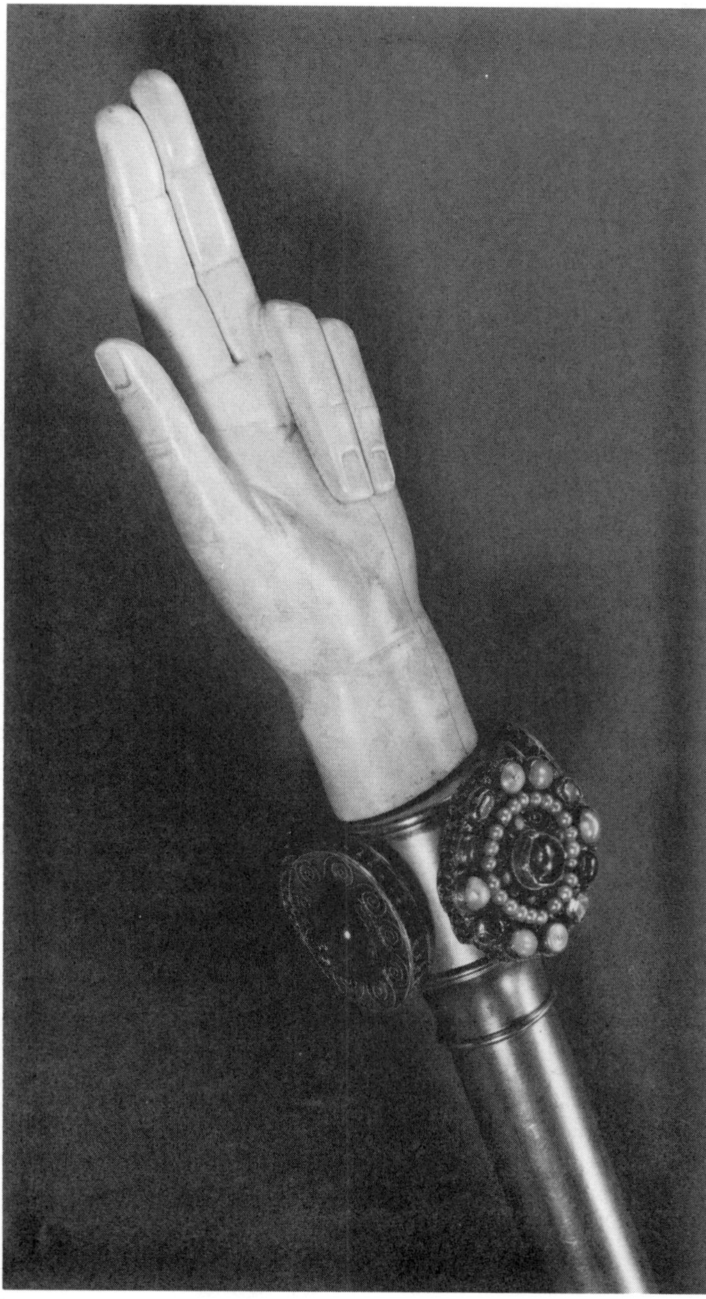

Figure 2.2 The sword "Joyeuse" and its scabbard

Figure 2.3 The scepter of Charles V

stones, and it is said to be the crown of Saint Charlemagne, Emperor and King; thus it is closed like an Imperial crown . . . the other crown was very sumptuous but it did not have a tiara" (Hervé Pinoteau 1970: 319–20). The interior cap, in the form of a tiara and crimson in color, comprises pearls stitched onto bands of gold.

The main crown is a circle of gold, consisting of

> four elements held together by four hinges . . . and each element was surmounted by a fleur-de-lis *issante* [upright] of very beautiful shape and spread wide. The stones were distributed uniformly, each rectangular element bearing a sapphire in each corner, a large balas-ruby and three emeralds; each fleur-de-lis had a balas-ruby on each petal and an emerald at the base, assuring the unity of the rectangular element with the fleur-de-lis by thus making the large balas-ruby to be surrounded by four emeralds.
>
> (Pinoteau 1972: 308)

What should be noted here is again the trinity of three stones in a set, surmounted by another, making 3 + 1 = 4, that is to say the *cosmological totality*. The four corners of the royal cloak are also likened to the four corners of the world, i.e., the cardinal points.[5]

Figure 2.4 The crown "of the Queen"

Pinoteau lays emphasis on the balas-ruby, known as an Alexandrian ruby, which is also to be found on the scepter of Charles V. He also stresses the simplicity of the stones' arrangement as compared with the other French crowns such as the *Couronne du sceau de Saint Louis*. In connection with the large ruby set at the front, Pinoteau points out that, "It is an international tradition to place a red stone in this position on the crown, since rubies evoke blood, whence sacrifice, a notion which has been intimately connected with that power for thousands of years" (Pinoteau 1972: 310). There is an analogy between the color red and blood of sacrifice which must be reiterated in the context of the sacred apparel, and also in the context of the very logic of the coronation.

As to the value of the precious stones, it can be seen that the crown contains a meaning – "The plan indicating a diminution of worth in a clockwise direction, the left-hand side nevertheless being endowed with a ruby of higher value than the one placed behind . . . (ibid.) – for the stones on the right were estimated at 12,600 livres, those on the back at 11,050 livres; and those on the left at 10,700 livres (ibid.). We see that the ruby placed at the front has higher worth

Figure 2.5 The Holy Ampulla

in this order of value than the other stones. In Pinoteau's opinion this is due to its symbolic significance as the "stone of sacrifice."

The arrangement of the crown, then, consists of sixteen rubies, sixteen sapphires, and sixteen emeralds divided up into four elements, each of twelve stones, which are of only three types. Once again we encounter the number three linked with the number four: the trinity linked with totality. The composition of the crown in four elements symbolized this totality.

It remains to establish the connection between the gold circle and the bonnet which together form the royal crown, and headgear of the characters featured in the lancet windows of the north rose of the cathedral at Chartres. These are David (or Solomon) and Aaron on one side and Melchizedek on the other: on one side the King wearing his golden crown; on the other the Priest, his head surmounted by the priestly tiara. The alliance of these two functions, the regal and the sacerdotal, is very clearly symbolized by this doubly prestigious crown with its tiara recalling the tiara of the Old Testament priests, and its circle of gold in imitation of the crown of the kings of Israel.

The sword is the one presented by Pope Leo III to Charlemagne which was used from the coronation of Philippe Auguste onwards, that is to say from 1179. This sword is known as St. Peter's, and it is also named "Joyeuse." The hilt, the guard, and the upper part of the scabbard are of solid gold enriched with gems; the scabbard, purple in colour, is garnished with pearls. This sword is worn in procession on the day of the coronation.

The acknowledgment: procession and celebration

The investiture is completed by the acknowledgment of the people, peers and Church. After benediction the King and Queen receive the kiss of peace from the Archbishop and from the peers. The Archbishop removes his miter to bestow his kiss (cf. the kiss of vassalage). He pronounces three times "Vivat rex in aeternum" (May the King live forever) and the King is presented in the company of two spiritual peers. (Before the twelfth century the people were asked for their explicit consent; later, a respectful silence takes its place.) Acclamations of "Vivat rex in aeternum" follow this presentation, while trumpets are sounded and coronation medals bearing the representation of the sovereign are thrown amongst the crowd.

It is related that on the occasion of the coronation of Charles VII bird-sellers released a multitude of little birds throughout the church. This happy note brings us to the ensuing festival and celebration which unite the members of the court and high church dignitaries at the palace while the crowd leaves the church in procession.

The King's peers play an essential role in the coronation. It is they who fetch the King from the palace, who stand around him during the enthronement and support his crown, and who bestow on him the kiss of peace and present him

Figure 2.6 Ordo of the consecration of the King (the consecration of Saint Louis). The unction of the King; the vesture with the royal cloak; the archbishop puts on the crown; the kiss of the peers.

to the people. The peers are twelve in number: six lords spiritual, and six lords temporal. The former are stationed to the left of the altar, the latter to the right. Hence the opposition left/right corresponds to ecclesiastics/laymen. The twelve peers are required to swear an oath of loyalty to the King which, at the coronation of Charles V, took the following form.

> You swear by your faith and oath and on the Holy Gospels that you will be good, loyal and trusty and obedient to the King of France our lord here present, to his heirs and successors, his body, his limbs, his heritage; the rights and nobility of the crown of France and of its sovereignty you will guard and defend before all and against all mortals. And you will give the King loyal and good counsel at all times he will ask it of you. And you will keep secret his counsel and all things which are to be kept secret for the good of the King and his kingdom.

This testimony of fidelity from the peers to the King is then renewed in the course of the coronation; the peers' oath is to the King what the King's oath is to the Archbishop. The two are directly related, the one reduplicating the other at a higher level.

The importance of the peers inheres in a form of "kinship" they have with the King. During the coronation there is an exchange between the King and his ecclesiastical and lay "family" in the context of the oath as in the context of the kiss of peace. The peers confirm the King's position by swearing to place their might and their substance at his service. The kiss of peace is the symbolic mark of the support they bring to the royal power, represented by their supporting the crown over his head at the moment of his enthronement.

The function of the eight stages of the royal coronation

The themes of death and rebirth, themes of the coronation

"The King sleeps," "the King retires," "on a ceremonial bed." In these successive phases of the King's preparation there is a prime preoccupation: to make the King disappear. He is physically absent while outside the cathedral the canons, the Archbishop in charge of the coronation, and the court nobles invited to the ceremony all await him. He is also separated from his wife, who nevertheless will be present during all the later stages. The ceremonial bed on which he sleeps is also the same as his death bed, for, prior to being placed in a coffin and buried, the deceased king is exposed for several days on a ceremonial bed. This withdrawal is necessitated by the state of sin in which the King stands at the time. And the King is no ordinary sinner. He bears the weight of the Christian community's sins – Frazer refers to it as "the burden of royalty." The death of the King in his role as knight is symbolized by the loss of his sword; his death as a simple human being is symbolized by sleep, absence, and the separation from his wife. As in all rites of passage he must be stripped of his previous state before he can accede to a fresh one.

The fallen knight no longer possesses escort or sword. He is removed from the sight of his people and his peers. It is the Archbishop who effects both the loss and replacement of the sword. Equally it is he who strips the King of his clothes for unction. The Archbishop performs as sacrificer, and the King as victim. What is sacrificed is the King's human nature before he is enthroned. The stripping of his clothes takes place in full view of the people; the King is symbolically reborn when he is invested with the regalia and when he receives the kiss and the crown which consecrate him. The following diagram represents the process in schematic form:

	death/	rebirth	
	human being/	King	
Stripping	"worldly state"/	"royal religion"	*investiture*
	ceremonial bed/	ceremonial costume	
	disarmed, without sword/	with sword, armed	

These two stages delineate what could be termed the political destiny of the King, who must first be presented as a victim before becoming a sacred sovereign through the intermediary of religious authority represented by the Archbishop.

Unction and communion: rites of sacrifice

Let us return to the origin of unction. For Saint Augustine, the sacraments are "the visible signs of things divine." The sacrament effects a transformation of the person who receives it: it consecrates him. Unction consecrates kings in this way as supra-temporal figures. "You will be transformed into another man," says Samuel to Saul at his unction (I Samuel 10:6). It raises man from his simple mortal state; the King shares the state of supra-temporality with the Bishop, God's representative on earth.

Having been baptized, Clovis was anointed and consecrated in his title of King with the Holy Chrism. Clovis' unction by Saint Rémi is an unction of confirmation made in the form of a cross on the forehead: "There [at Reims] the Frankish nation and its King . . . were worthy to be washed in the holy water of baptism and enriched by the holy grace of the septiform spirit" (Flodoart, *Historia Remensis Ecclesiae*, Book I, Chap. XIX).

In the Western Church up to the eighth century unction was given after baptism. There was, then, one unction given by the priest as a complement to baptism and another given on the forehead with the Holy Chrism by the bishop at the time of confirmation. When priests baptize, they are permitted to anoint the baptized with the chrism, but on the head, not on the forehead, which is the prerogative of bishops when they transmit the Holy Spirit. Until the thirteenth century the Church's position was very strict: ordination figured among the seven sacraments, but not unction.[6]

Jean Golein's text on the coronation of Charles V in 1364 serves to illustrate the point: "When the King strips himself, it is with the significance of relinquishing his previous worldly state in order to take up that of religion . . . and if he take it up with such devotion as must be, I hold that he is cleansed of his sins as much as he who newly enters into the proven faith." There is a true change of state in this context: the King passes from the secular world into the religious and sacred world symbolized by the sacraments.

Saint Bernard, who takes up the subject in his *De precepto et dispentatione*, affirms that "just as at baptism sinners are pardoned so are they on entry into religion." Unction is thus a ritual of regeneration comparable to the sacraments of baptism and ordination of which it is part. Bloch (1961: 196ff.) suggested that unction, priestly and royal brought face to face, served finally to effect a more extended use of unction; to justify and to legitimize, as well as to purify.

The usurper who wishes to legitimize his power, obtained by violence, subjects himself to baptismal unction, which effaces the murder and bestows on him the purity of the newly baptized. The benefit deriving from the fusion between baptismal and royal unction can clearly be seen; the fusion is rendered necessary by the imperatives of power, namely to gain the guarantee of legitimacy accorded by a sacred right.

It is not necessary to follow literally Frazer's interpretation of kingship, drawn from the exotic examples of the rule of succession to the priesthood at Nemi, to concede that, by the fact of his succession to the throne, the successor, quite regardless of the instances where it was literally the case, assumed a kind of symbolic responsibility for the demise of his predecessor. Hence before the coronation the new king appears in the guise of his murderer.[7] It is his state of guilt which justifies the sacrament of unction. The very function of royalty embraces putting to death. In this sense the whole ceremony can be understood as a purification to place the King in a state of grace equivalent to that of the newly baptized, and pardon him for the acts of violence he has committed and will commit in the future.

In the ceremony the royal *ordines* (ceremonial rules) come together once again with those of the priesthood: apart from the three superimposed robes, after unction King Charles V put on gloves as bishops are accustomed to do at their consecration. Despite the efforts of Innocent III, Alexander IV, and the curia which followed them, who attempted to abolish the use of holy oil for the kings, in both England and France its use continued. Moreover, this use of holy oil is substantially complemented in the course of the coronation. For, during the coronation mass, the King performs a priestly office, that of a sub-deacon: he is allowed to assume the ceremonial conduct of a priest, and a place is accorded him within the hierarchy. This is corroborated by the lesson which the prior of cardinal deacons reads over him, for it consists of the customary litany employed in the ordination of sub-deacons. The kiss of peace bestowed by the Archbishop is that normally received by cardinal deacons. Certain kings wished actually to use their role of deacon; Charles IV, for example, read in the church on Christmas day the seventh lesson of the matins, taken from the gospel used in the midnight mass.

In the light of these various attributions it could be said that the King has equally a priestly role: it is a king-priest whom the Archbishop consecrates at the enthronement. This priestly role in itself confirms the King's religious position, and it is particularly evident in the ritual of communion which constitutes an effective participation of the King in the sacrifice of the mass.

In *Le Traité du normand anonyme*, written by a cleric in the late Middle Ages, account is taken of the mixed nature of the King as a composite person or "persona mixta." This persona represents the union of the King's secular powers and the Archbishop's spiritual ones: a union of the royal blade and the priestly cross, symbols of the two powers. The central idea of the treatise lies in the identification of the King with Christ: *christomimetes*, an epithet taken from Byzantine theology where the King is constantly likened to a living god.

Beyond the King's spiritual character, granted through the clericalization of his office, there exists an impersonal and immortal body.

> We thus have to recognize [in the King] a twin person, one descending from nature, the other from grace . . . one through which, by the condition of nature, he conformed with other men; another through which, by the eminence of [his] deification and by the power of the sacrament [of consecration] he excelled all others. Concerning one personality he was, by nature, an individual man; concerning his other personality, he was, by nature, a Christus, that is a god man. (Kantorowicz 1957: 46)

The nature/grace antithesis is the vehicle of Christian deification in the Middle Ages: grace does not serve only to palliate the natural weakness of human nature, but renders man more suited to participate in the divine nature itself. "The power of the King is the power of God. This power, namely, is God's by nature, and the King's by grace. Hence the King too is God and Christ, but by grace; and whatsoever he does, he does not simply as a man, but as one who has become God and Christ by grace"[8] (ibid.: 48). It is this double essence of royal power which makes of the King a twin person: "the King is a twinned being, human and divine, just like the god-man . . . the terrestial King is not, he becomes a twin personality through his anointment and consecration" (ibid.: 49).

At the time of communion there is a repetition of the sacrifice of the Last Supper. The King, in the midst of his twelve peers and taking communion in both kinds, is the central figure of the rite. The hymns and prayers at the time of communion signify sacrifice: "Blood, mystery of faith which for you and for many has been spilt in remission of your sins." Christ is conceived of as a human victim, offered for the redemption of the world, and it is this sacrifice which the mass daily reenacts in symbolic form. The King who brings his offering of bread, of wine, and of gold makes thereby his contribution to the sacrifice performed by the Archbishop. By furnishing the sacrificial material he confirms the symbolic link which he maintains with the original victim, that is to say, with Christ.

In the prayers which accompany communion the following oppositions are encountered:

Heaven/Earth
Christ /Soul
King /People

The rite of communion exalts the mystical union between these terms which are opposed two by two. The idea of sacrifice is also encountered in the symbolism of the Lamb, a victim chosen in substitution for the original victim and traditionally sacrificed at Easter in memory of Christ's rebirth. It is at Easter that communion is made obligatory for all Christians in recollection of the founding sacrifice of the Christian religion.

The Cene, which the Eucharist reenacts, is the last meal taken in common with the apostles, when Christ demands of each of them the oath of loyalty and sworn faith that is equally to be encountered in the institutions of feudalism and in the coronation (for example, in the peers' oath). Royal communion thus accentuates the parallelism between the King's destiny and Christ's. Like unction, it prepares the King for his "entry into royal religion"; it establishes grace as the active principle of divine incarnation. For it is by the intermediary of communion that the King participates in the divine nature, as we have seen in *Le Traité du normand anonyme*. Without these rituals which allow the King to acquire further grace no sacralization would be possible.

This superabundance of grace privileges the King in his relationship with men of honor: the peers and the people. It allows him to stand out from both, by encompassing them and by giving them a part to play in the ceremony itself: the peers become the witnesses, the people the spectators.

The sword of justice: initiation to power

The initiation of the King to power is effected through the return of his sword. It has been seen that the repossession of the sword goes through several stages and that it corresponds with the King's initiation as a knight. The sword is successively given to the King, then taken from him, then returned to him, then carried in procession before him at the close of the ceremony. It symbolizes the role of royalty, which goes with the right to exercise power in the name of the people. A prince reigns in accordance with the principles of law, that is to say, according to Kantorowicz, in pursuance of the duties with which he is charged. The notion of divine right has its roots in this medieval conception of sovereignty. There is an antinomy between the King's public person (*persona publica*) and his private will (*privata voluntas*) which is to be encountered in the following quotation from John of Salisbury: "That the prince, although he be not bound by the ties of law, is yet Law's servant as well as that of equity: that he bears a public person *and that he sheds blood without guilt*" (John of Salisbury, quoted in Kantorowicz 1957: 95, my emphasis). Royal Justice is thus based on this double principle, namely, that the King is simultaneously the master and slave of equity, and able to spill blood without incurring guilt. He simultaneously represents the law and the law's transgression whether by murder or any other crime.

The King as a figure of justice is central to all medieval treatises on law: the idea of justice is tied to the law and at the same time is above the law since it represents the ends of all law (Kantorowicz 1957: 96). This ambiguity follows from the double nature of the King, which we have just seen. Thus, through the image of the King as mediator, we can pass from the liturgical into the juridical sphere.

This ambiguity is to be encountered in the law itself, sometimes qualified as *lex digna* or as *lex regia*; for these are two contradictory laws of the Roman legal system, one making the King a slave to the law, though not being subject to it (*lex digna*), the other releasing him from all constraint (*lex regia*) and making him sole master of the law.

The sword of justice handed to the King is referred to as *pro fide*, i.e., "for the faith." Kantorowicz connects *fides* and *fidelitas* in the sense of feudalism containing a double connection: first of loyalty towards a lord; secondly of faith in respect of God. The medieval Christian leaves for the Crusades with the idea that, should he die, he will inhabit a city in another world, and he will be a subject in a kingdom. His model is Christ, who has sacrificed himself *pro fide*.[9]

The peers' kiss: the material exhibition of royal power

The peers express their fidelity to the King by giving him the kiss of peace at the moment of benediction.

> In ancient times there were only twelve peers in France, six ecclesiastics and six laymen . . . These ancient peers placed themselves as they had to . . . in judgment in the presence of the King . . . Once the crown was placed on the King's head by the Archbishop, all the peers, both clerical and lay, had to put their hands to it and support it from both sides. Then the archbishop and the peers who were holding up the crown had to lead the King thus adorned to the chair which was prepared for him . . . And the Archbishop out of reverence had to kiss the King in such manner installed on his seat. And afterwards the bishops and the lay peers, who were holding the crown [had to kiss him] putting on the sword as it has been previously said. (*Le Livre du sacre des rois*)

The peers are placed around the King like a court of justice. Here we touch on the origin of the peers:

In all [court cases], civic and military, the Franks had twelve witnesses, eight from the paternal side, four from the maternal side, all of whom had to be truly Salic Franks. At the coronations of kings the eight witnesses from the paternal side were represented by the principal lords of the kingdom. The four relatives from the maternal side were represented by four bishops . . .

(Note from *Cérémonial du sacre des rois de France*, 1775)

The assembling of twelve relatives around the King is needed in order to constitute a sovereign group of twelve, the minimum number of witnesses to establish a court of justice.

The ecclesiastical peers are placed on the King's left, the side on which the sword is worn and the hand of justice is carried. The secular peers are on the right, and the scepter, too, is held in the right hand. It is the symbol of the King's fief, used also in the investiture of prelates. The Queen is placed on the left on the same side as the ecclesiastical peers and as the maternal relatives in ancient judicial proceedings.

Marc Bloch explains, in *La Société féodale* (1968), that feudal ties are supplementary to the ties of kinship and are, in a sense, assimilated to them because of the inadequacy of the ties of blood. Feudalism exists to the degree that it is founded on the different groups formed around the King. Feudal society is hierarchical, and Bloch underscores the economic subjection of the "humble" class to a double oligarchy, that of prelates or monks, charged with "rendering heaven propitious," and that of warriors, i.e., a class of seigneurs and professional soldiers. In the eleventh century the social link, in fact the subordination, of the individual to public power is expressed in a quite positive way: the peers surround the King and constitute his general staff, and do not directly put royal power into question save in a case of conflict grave enough to shake the feudal edifice to its foundation.

The defense of privilege is the cohesive factor binding together all the members of the sovereign group formed around the King. This is why kinship, real or fictive, plays so important a role in the sovereign group's domination; kinship is the framework which gives strength to the fidelity between the various elements of the group.

We can now see how a space is established around the King which has its own dimensions (left and right, behind and in front, objects and persons being aligned, each on a well-defined axis with its place determined in advance: everything happens as if, by means of the ritual, the history of coronation were reenacted on each occasion).

Moreover, in the rites of the royal funeral which precedes the coronation it becomes clear that the same order and logic, applied with the same care, provide the meaning given to the objects and beings which people this space.

The funeral

The King is dead, long live the King!

Funerals are extremely ostentatious ceremonies, and they also include moments of intense revelry, periods of profound mourning, and, once again, of triumph. The preparation, sacred vesture, unction, investiture, victory, procession, and celebration once more bear witness to the fact that "the King never dies."

Two fundamental features provide the clue to this ritual: on the one hand the creation of an effigy made in the image of the King (Kantorowicz describes the King's entry into the church as a "triumphal *adventus*" rather than a funerary procession); and on the other hand, the destiny of the royal honors, or regalia.

The preparation

When the King dies, his body remains for several days in the funerary chamber lying in state on a ceremonial bed with his face uncovered. The body thus arranged is the object of everyone's veneration. During this same period, a wax effigy modeled from life is made to receive the honors due to him. The corpse is then placed in a lead coffin under the ceremonial bed on which the effigy lies.

The body of the King is embalmed immediately after his death. It remains in a coffin for the forty days during which the effigy placed above it is displayed on the ceremonial bed (according to Rousselon, this was the procedure for King Charles VI, Charlemagne, and Saint Louis).

The effigy is dressed in the King's ceremonial robes: shirt, camisole, tunic, and cloak. On its head is a bonnet of dark red velvet and the royal crown. The head rests on a cushion, also of red velvet, with the King's scepter on the right and the hand of justice on the left. Facing it a richly adorned altar is set up. At the effigy's feet stand two heralds whose duty it is to offer holy water to the princes and to those who have been granted the honor of sprinkling the corpse. During the period of the effigy's display, everything takes place as if the King were still living.

The creation of effigies takes us back to the period of the Roman and Byzantine emperors who had likenesses made of themselves to mark their triumph over death, and Berthevin connects it with the pagan ceremony of apotheosis. The historians Dio Cassius and Herodian give the following account of the deification of the Emperor Pertinax. His artistically prepared effigy was laid out on a triumphal couch. It displayed the *princeps* dressed in the clothes and all the ornaments of imperial dignity. For several days the doctors made their customary attendance, taking the dummy's pulse and issuing reports on the state of its health. Meals and meetings took place as usual, and it was considered that the *princeps* participated in them.

It is thus in imitation of these rites that medieval monarchs employed effigies. The effigy has, however, a quite specific meaning here. It represents the King as a double person, that is to say, it symbolizes the deceased monarch and also the function of royalty which itself never dies. There are, therefore, two royal persons, each in his own context within the same hall: a state room where the effigy reposes in full ceremonial garb, the depository of the royal attributes where all come to pay their respects, and at the same time where the corpse lies dressed in black.

Once the forty days are over, the state room is transformed into a mourning chamber, and the effigy is replaced by the corpse which is placed on a trestle in its coffin; it is covered by a black cloth with a cross. Over this is another cloth of gold with a white cross and the arms of France.

At the head of the coffin there is a pillow of cloth of gold to receive the royal crown; on the right is the scepter, and on the left the hand of justice; at the foot of the bier stand the heralds dressed in coats of arms; the catafalque is draped with a banner and two altars are set up at the end of the room. A few days before the removal of the corpse, the King's successor comes to kneel beside it. He prays and sprinkles it with holy water; he leaves his royal cloak of red in the hands of the heralds at arms charged with the surveillance of the King's body.

Loss and restoration of honors: procession and celebration

The death of the King is announced throughout the whole city by the voices of twenty-four town criers, and his inhumation follows.

The body is taken out of the candle-lit chamber where it has lain, and it is taken in procession to Notre-Dame de Paris where the vigil service takes place, that is to say, a ceremony performed the evening before the actual day of the funeral, and in the presence of the corpse alone without the effigy. This ritual incorporates the same regalia and follows the same course as the one which will be celebrated next day at Saint-Denis. The coffin is preceded by the town criers of Paris and the poor. It is only after this that the effigy is placed on a litter with the royal attributes in preparation for the proceedings of the following day.

During the entirety of the vigil service the coffin containing the corpse is carried by four lords. On the right-hand side of the corpse are the princes of the blood, on the left the cardinals with the celebrant priest in the middle. When the service is concluded at Notre-Dame, the lords who carry the banner, the knight's standard, the pennant, and the guidon will leave them planted there until the next day. These insignia will be carried and taken up by these same lords when the corpse leaves, surmounted by the effigy which will then be uncovered.

On the appointed day all the corporations of the city of Paris leave in the direction of Notre-Dame. Each one sprinkles the coffin and the effigy with holy water. The ceremonial bed is carried by the gentlemen of the bed-chamber. The effigy is laid out on this bed, the crown on its head, the scepter and the hand of justice in its hands. The Grand Equerry, bearer of the royal sword, walks before the ceremonial bed. He is preceded by a riderless horse whose black caparison is held by four valets dressed in black. The effigy is surrounded by "the court of parliament," that is to say the judges in their robes, and it is the four presidents of this court who hold the four corners of the pall. On one side of the effigy is a throng of archers with their captains, the monks, the poor, the squires, the twenty-four criers, the constables of the watch, the advocates, and the provost of Paris and his aides. On the other side are forty bishops and archbishops, dressed in black copes and with white miters. Before entering the church, the convoy halts and the Abbé of Saint-Denis addresses the officiant priest asking him three times if it is truly the body of the dead King and, on his affirmation, allows it entry.

Mass is said at Notre-Dame and at Saint-Denis; at the moment of the offertory one of the masters of ceremonies goes to fetch the first "prince of grand mourning" to lead him to the offering. When the prince is on the point of kissing the paten, he takes from the hand of a King-of-Arms a white wax candle on which are fixed four gold *écus*. The other princes of the blood follow suit. While the *De Profundis*, said in honor of the King, is drawing to a close, the masters of ceremonies go to gather up the royal regalia and hand them over to three dukes. When the body has been placed in the burial vault, the King-of-Arms cries, "Heralds-at-Arms of France, come perform your duties!" The heralds and sergeants-at-arms throw down their tunics and their maces. The herald-at-arms of the house of Orleans descends into the vault to arrange on the coffin all the items of the regalia: the insignia of the various corps of the royal guard are then thrown down into the vault, then the King's spurs, gloves, coat of arms, helmet, pennant, and shield (everything which constitutes the King's apparel as a knight), finally the hand of justice, the scepter, and the crown. The Masters of the Household throw down their batons. The Grand Equerry, who holds the sword by its hilt, inserts the point alone down into the vault, and then withdraws it immediately to hand it over to the new King.

The Grand Master of Ceremonies puts the end of his baton into the vault and says in a low voice, "The King is dead," and then, "Let us pray for the peace of his soul." Each person kneels and prays in silence; after a few minutes the Grand Master cries, "Long live the King!" pronouncing the name of the successor. The successor then receives at this point the sword as the sign of his succession. The King-of-Arms receives on a cushion of black velvet the regalia which have been retrieved from the vault, i.e., the hand of justice, the scepter, and the crown. Once the banner has been raised again and the Grand Master has taken up his baton, festive music sounds throughout the church and everyone leaves to prepare himself for the celebration which must follow. All the heralds-at-arms again take up their insignia; the first steward says to all of them, breaking his baton of command, "Sirs, our master is dead and let each man look to himself for the house is broken . . . " These words announce to all the officers of the King's household that they are relieved of their duties.

At the conclusion of the funeral the effigy is removed and burned like a common bundle of straw, having fulfilled its function.

The two bodies of the King

The most remarkable feature of this ceremony is the presence of the triumphant effigy, dressed in full ceremonial robes, entrusted with the royal attributes, and placed beside the King's body which is recumbent with hands clasped and dressed in mourning.[10] This was clearly illustrated in the funeral of Charles VIII in 1498. The cortège conveying the corpse alone passed by with doleful show: everything was draped in black, flags were furled, the sword of the kingdom was sheathed, and the other emblems of royalty were covered over.

Upon entering Paris, the effigy was placed on top of the coffin, and then, as Kantorowicz describes, the whole mood completely changed. The sword, naked, preceded the effigy, adorned with the regalia.

Amidst the royal cortège it is notable that only the effigy and the high dignitaries of the court of justice are permitted not to wear mourning, but on the contrary display their full ceremonial robes which in a fashion corresponds with their role. Judges, says Kantorowicz, have a privileged relationship with the King: they are *pars corporis principitis*, a part of the royal body. They are associated with royal dignity and in a way share in it; it is they who are empowered to ensure that the ordinances pronounced by the King are respected, and by this means they constitute an integral part of the King's juridical function. Moreover, their judicial dress is related to that of the King: it consists of a red robe and a fibula securing a cloak at the shoulder (the fibula is frequently mentioned by medieval jurists as one of the insignia of imperial majesty, together with the purple, the scepter, and the diadem).

Let me quote Flavin (1921) in connection with this detail: "The justice and even the authority of parliaments is considered always to persist in this kingdom, even if the King is dead . . . or absent" (XIII, c88, paragraph 10, p. 1, 181). "Parliament as a body never wears mourning, but the court attends funeral ceremonies as a body in its red robes, and walks with the effigy of the King, which is on a bed, as if accompanying the King on his bed of justice" (paragraph 29, p. 1, 186). The association of the ceremonial bed and the bed of justice bears witness to the majesty of the King's *persona*.

Certain expressions which belong to the funeral ceremony cannot be understood except by reference to what happens to the corresponding regalia; thus "the crown never dies" or "the banner never dies" must be related to the movement of the other symbols of majesty. The regalia are first assigned to the effigy, then to the King's corpse up to the moment of its interment, and finally to the succeeding King after they have been brought up from the vault. These expressions demonstrate the continuity which overrides the break caused by the royal death and the succession which immediately takes up the attributes of power. Everything happens as if an attempt were being made throughout the ceremony to diminish the import of the event represented by the King's death and the fact that a successor to him must be appointed: the forty days' honors paid to the effigy bear the same sense, covering the arrangements made for the succession and the preparation of the body of the King for its integration into its new abode. The body is dressed, it is cared for, its double is created, it is consecrated, it is blessed and, just as at the time of investiture, it is surrounded by the peers, the judges, its family, and it retains the armaments of knighthood as a proof of the courage shown in the combats which had to be faced during the King's lifetime. All these operations contribute not just to making it a simple relic, but to its preparation for a higher realm and its constitution as an ancestor.

The funeral is in itself a ritual and political system which transcends the question of simple personal power; it guarantees by the precision of its rites and the symbols which it uses the affirmation of the vigor of the institutions which constitute royalty. This is not limited to the person of the King himself but through his person extends to the whole system and the whole group which exercises sovereignty.

Yet the royal corpse has itself a sacred value: it is by *sancto corpore* that miracles are performed. Even dead, the King has a thaumaturgical role, and it is through the King's role as a miracle-maker and healer that the sacred nature of royalty can most nearly be approached.

Royal justice and thaumaturgy

I have shown that there is a correspondence between the cycles through which the sword passes at coronations and at funerals. The successive relinquishings of the sword which finally result in its solemn return signify the King's laying aside of his human personality, a necessary process for his confirmation as a cosmic power (in the coronation) and his entry into a higher realm (in the funeral). To attain the status of a cosmic persona and a being of the higher world the King must divest himself of his human appearance and be reborn a god. Death and rebirth are the two rites through which the King must pass to confirm his mastery over nature. Indeed it is towards this end that all the stages of both ceremonies are directed: to render the King sacred in such a way that his divine nature is irrefutably established.

The significance of the sword as something presented to the newly elected King which belonged to the previous King is something more than as a mere symbol of temporal power. It is the *instrument of sacrifice*, and in accepting it he not only accepts the attributes and duties of his predecessor but also, with the responsibility for replacing him, the symbol of murder, the guilt of the initial murder[11] on which all dynasties are conceived to have been founded, the origin of every establishment of power. It is this murder whose trace unction must efface in the same way as baptismal unction effaces the sin of Adam. Herein lies the importance of the royal sacrament whose value has always been denied by the Church.

The King's power of life and death over his subjects is confirmed by his role as a healer, for he also has power over sickness and is capable of performing miracles. This gift of healing is something which is transmitted by heredity within the royal families of France and England and which can be lost as a result of some grave crime in the eyes of the Church (adulterous marriage, for example).

It is said that on his coronation day the King can cure scrofula by a simple laying-on of hands.[12] The French phrase "toucher des écrouelles" which describes this gift has its English equivalent in the term "King's evil" (cf. Frazer, *The Golden Bough*, p. 118). Again, according to Frazer, "scrofula was regarded . . . as a result of eating with tabooed hands . . . " (ibid.: 269). The King is thus likened to a sorcerer or healer. In *Les Rois thaumaturges* Bloch (1961: 245) gives a description of the royal sign which coincides with the healer's sign: "It is a quite widely held opinion amongst the common people that individuals gifted with particular magical powers, and especially healers, brought with them into the world at their birth a distinctive mark on their body, an index of their talent and sometimes of their illustrious origin." Generally this little sign appeared on one shoulder and distinguished whoever bore it as possessing certain magical powers. Nogaret says in his "Mémoire justificatif de 1310" (p. 519) in connection with the laying-on of hands, that: "God by the hands of the King performs evident miracles for the benefit of the sick." This healing power which allows the King to increase his popularity amongst the people who firmly believe in it gives medieval royalty its distinctiveness: in Bloch's view monarchical power is *sui generis* within the feudal world, and of a different order from that of other forms of domination. Royalty is "feudalized," that is to say the King is present at all levels, as lord, and equally as priest and magician. Through the appropriation of channels of feudalism the person of the King and his mystical virtue are imposed on the people. The sacred character of the monarchy is not intimately connected to heredity; there remain, even within the coronation ceremony itself, traces of the sort of election which was practised during the first period of the Frankish monarchy: two bishops, at the time of the coronation, present the King to the people for their assent; the people show their agreement by acclamation. From the time of the coronation of Louis XI in 1461, this presentation no longer occurs as part of the ceremony, but takes place after it.

To summarize, the coronation legitimizes the King's sacerdotal and royal functions: it allows him to obtain the consent on the one hand of the Church, through whose intermediary everything is done, and on the other of the people, by whose faith he is consecrated as a miracle worker, and of his peers, lords, and bishops who constitute the sovereign group bound to the King by ties of fictive if not real kinship. The coronation confirms the King as a sovereign *by grace of God*, and hence in his definition as a sacred personage superior to other mortals and possessing a mastery over nature – "a Phoenix King reborn from the ashes," according to a fifteenth-century manuscript (quoted by Kantorowicz 1957: 414). He also has power to cure the sick, to make crops grow, and to increase fertility; through his justice he gives order to the world and guarantees peace throughout his kingdom; his body itself becomes sacred after death, and the funeral ceremony consecrates him as a being of the higher realm capable of continuing his work through the person of his successor.[13]

Notes

1 "The Saturday before the Sunday on which the King is to be consecrated and crowned, after the singing of Compline, the church must be given to be protected by such guards as are established by the King, together with the church's own guards . . . When Matins are rung, the King's guards must be drawn up to protect the entrance to the church . . . And the other entrances to the church must be well and firmly closed and garrisoned." (*Le Livre du sacre des rois*, chapter on "Regulations" concerned with the preparation of Charles V for his coronation.)

2 "And early in the evening the King must come to the said church to offer prayer and keep vigil there some time in prayer if he wishes" (ibid.).

3 The same association is made, inverted, in Jamous' chapter in this volume on the Rif where, in the symbolism of marriage, the bridegroom figures as the Sultan. (Eds.)

4 According to the inventory of the treasury made in 1534, and again in 1634, it weighs 16 marcs 4 ounces. The marc, or Paris half-pound, weighs 244.75 grams. So the weight of the whole is over 8 pounds, being heavier than the crown of St. Edward, which weighs 4 pounds 15 ounces, and the Imperial crown of Vienna, which dates from Otto I and weighs 12 Viennese marcs 3 half-ounces.

5 We shall encounter the royal cloak again in the ritual in the chapel in Versailles, described in the next essay by Leroy Ladurie. (Eds.)

6 According to Renan (1870), "France had created an eighth sacrament . . . which was administered only at Reims, the sacrament of royalty."

7 See the chapter by Jamous in this volume where it is explained that the role of succession to the Sultanate requires that he should shed blood before acceding to legitimate authority. (Eds.)

8 In the above two quotations Kantorowicz is citing *Le Traité du normand anonyme*, published in 1899 by Heinrich Böhmer.

9 *La Chanson de Roland*, last canto, "He who embarks on the Crusade who dies in the field, will enter into the divine kingdom and will dwell with the angels."

10 In Kantorowicz's view, the effigy represents the *persona ficta*, that is to say, the dignity of royalty as opposed to the real person or corpse of the King which represents

his mortal remains. His distinction is fundamental and, although he refers to England rather than France, the differences are few in the medieval period; the English monarchy was first Norman and then Angevin. But Kantorowicz was concerned with theology and the juridical theory implied by it. This essay deals with the ritual symbolism explicating his thesis.

11 Susan Drucker-Brown's analysis of the royal rituals of the Mamprusi comes to a similar conclusion: the King must die alone as only a murderer is condemned to die, lest his spirit take vengeance on those present. (Eds.)

12 Scrofula was a localized form of tuberculosis taking the appearance of a tumor or ganglion usually on the neck.

13 There was an exhibition in the Louvre under the title *Regalia: les instruments du sacre des rois de France. Les Honneurs de Charlemagne*, 14 Oct. 1987–11 Jan. 1988. Its catalogue is of the greatest interest to those concerned in the subject. (Eds.)

References

Bakhtine, Mikhail 1970. *L'Oeuvre de François Rabelais et la culture populaire au moyen âge et sous la renaissance*, Paris, Gallimard

Bataille, Georges 1965. *Le Procès de Gilles de Rais*, Paris, Pauvert

Baudrillart, Alfred (Monseigneur) 1929. *Le Sacre de Charles VII et le triomphe de Jeanne d'Arc*. Discours prononcé dans le cathédrale de Reims à l'occasion de 5ème centenaire, le 17 juillet 1929, Paris, Editions Spes

Berthevin, Jules 1825. *Recherches historiques sur les derniers jours des rois de France, leurs funérailles, leurs tombeaux, suivies d'une notice sur st Denis, le sacre des rois et leur couronnement*, Paris, Louis

Bloch, Marc 1961. *Les rois thaumaturges*, Paris, Armand Colin

Bloch, Marc 1968. *La Société féodale*, Paris, Albin Michel

Brehier, Emile 1949. *La Philosophie du moyen âge*, Paris, Albin Michel

Brehier, Louis 1956. *Les Institutions de l'empire byzantin*, Paris, Payot

Cérémonial du sacre des rois de France 1775. Paris, Manuscrit de la bibliothèque nationale

Certeau, Michel de 1970. *Les Possédés de Loudun*, Paris, Julliard

Chadwick, Henry 1971. *The Early Church*, London, Penguin

Drucker-Brown, Susan 1975. *Ritual Aspects of the Mamprusi Kingship*, African Social Research Documents 8, African Studies Centre, Cambridge University Press

Dumont, Louis 1967. *Homo hierarchicus*, Paris, Gallimard

Flavin, Bernard de la Roche 1921. *Treize livres des parlemens de France*, Geneva

Frankfort, Henri 1971. *Kingship and the Gods*, Chicago, Chicago University Press

Frazer, James George 1980. *The Golden Bough*, abridged edition, London, Macmillan

Freud, Sigmund 1950. *Totem and Taboo*, London, Routledge and Kegan Paul

Freud, Sigmund, 1985. *Moses and Monotheism*, Harmondsworth, Penguin

Gaborit-Chopin, Danielle 1987. *Regalia: les instruments du sacre des rois de France. Les "honneurs de Charlemagne"*, Paris, Edition de la réunion des musées nationaux

Godefroy, Théodore 1619. *Le Cérémonial français, ou description des cérémonies, rangs et séances observées aux couronnements, entrées et enterrements des rois et reines de France*, Paris, Pacard

Golein, Jean 1372. Traité du sacre, in Rational des divins offices, manuscrit de la Bibliothèque Nationale 437, Fol. 2 V° Coll

Hocart, A. M. 1969. *Kingship*, Oxford, Oxford University Press

Hocart, A. M. 1970. *Kings and Councillors*, Oxford, Oxford University Press
Kantorowicz, Ernst H. 1957. *The King's two bodies*, Princeton, Princeton University Press
La Chanson de Roland 1971. Paris, Gallimard
Le Livre du sacre des rois (Charles V) 1896. Extrait du bulletin historique et philologique, BN 8° Q 1054
Machiavelli, Nicolas 1977. *The Prince*, trans. and ed. by R. M. Adams, New York, Norton
Nepote, Jacques 1972. *Les Sept phases du couronnement*, Nanterre
Nogaret, F. Mémoire justificatif de 1310, in P. Dupuy, *Histoire du différend d'entre le pape Boniface VIII et Philippe le Bel en 1655*
Ostrogorsky, Georges 1956. *Histoire de l'état byzantin*, Paris, Payot
Pange, Jean de 1949. *Le Roi très chrétien*, Paris, Fayard
Pange, Jean de 1955. Le Sacrement de Reims, in Recueils de travaux offerts à Mr Clovis Burnel, T II, pp. 349, 354
Pinoteau, Hervé (Baron de) 1970. *Une Représentation du sacre de Claude de France (1527) et quelques considérations préliminaires sur les insignes de Charlemagne*, Madrid, Hidalguia
Pinoteau, Hervé (Baron de) 1972. *L'Ancienne couronne française dite de Charlemagne*
Renan, Ernest 1870. *La Monarchie constitutionnelle en France*, Paris
Rousselon 1775. *Cérémonies usitées à la mort des rois de France*, BN Lb 48 2686
Southern, R. W. 1972. *Western Society and the Church in the Middle Ages*, London, Penguin
Troeltsch, Ernst 1949. *The Social Teaching of the Christian Church*, London, George Allen and Unwin
Yamaguchi, Masao 1972. *La Royauté comme système de mythe*. Diogène 77

3

The court surrounds the King: Louis XIV, the Palatine Princess, and Saint-Simon

EMMANUEL LEROY LADURIE
(translated by Roger Just)

The problem of rank in Louis XIV's court is well known, particularly thanks to Saint-Simon's memoirs[1] and the thesis of J.-P. Labatut. While the letters of the Palatine Princess, known as "Madame," are less well written than the Little Duke's text (it is true that they lose some of their savor in the translation from German into French), the Palatine's correspondence does have the advantage over the Memoirs of being at times more condensed, of saying rather more in fewer words. This may be judged by the following letter, written at Versailles on December 27, 1713:

> . . . my son is grandson of France. Grandsons of France are ranked above the princes of the blood; it is true that they do not have as many privileges as the children of France, but they have many more than the princes of the blood. Thus, my son eats at the king's table, whereas princes of the blood do not eat there. He has never borne the title of first prince of the blood, for he is not a first prince of the blood, but grandson of France. This is why he is addressed as "Royal Highness." His son on the other hand, first prince of the blood, is addressed as "Serene Highness" and not as "Royal Highness." He does not attend the king at the levee or the couchee; he is present only for important ceremonies, when the entire family dines with the King; thus he has no right to the privileges enjoyed by his father, such as the "studded carriage",[2] First Squire, First Almoner, etc.; his officers may not and should not serve him in the King's presence; he maintains no retainers within the palace, and a hundred other things of this sort separate him from my son. I must have made an error in writing, for my son has never been a prince of the blood. The King has indeed accorded the rank of prince of the blood to the Duc du Maine[3] and to his sons and to his brother, but they remain outranked by all the other princes and princesses of the blood. This is true to the extent that, in her own household, the Duc du Maine's wife is seated above her husband and has precedence over him in all matters; when a contract is to be signed, she signs in virtue of her rank by birth, whereas he may add his name only after those of all the other princes and princesses of the blood. He is thus ranked far below my son: between them come all the princes of the blood. As regards my own position, it remains unchangeable. Were the King to have a daughter, she would be called "Madame," and I would be addressed as "Madame, Duchesse d'Orléans." My son's wife is addressed as "Madame, *la Duchesse* d'Orléans." This use of the article indicates that she is not a child or daughter of France, but only a

granddaughter. One must be familiar with this court in order to distinguish all these differences correctly. Monsieur le Dauphin acted in a gravely unjust manner towards my son's daughter when he decided that she would come after the married princesses of the blood. However, it is quite certain that her place is first, her brother being the first prince of the blood. But since Madame la Duchesse was the favorite of the first Dauphin, she was able to make him do whatever she wished, and the King did everything Monsieur le Dauphin asked of him. Had the latter lived, the princes of the blood would have gone far . . .

This is an extremely rich account. First, it gives the whole scale of different ranks – sons or children of France, grandsons of France, princes of the blood and, lastly, royal bastards (the Duc de Maine, etc.) – precedence of the bastards over the rank and file of dukes and peers being what so violently embitters Saint-Simon. This text also describes the material or symbolic signs which distinguish the different ranks: whether or not one eats with the King, how much time one spends with him, whether or not one's officers serve their master in the King's presence, whether or not one is entitled to a studded coach, etc. At the same time some indication is given of the intrigues which did not overturn the fundamental hierarchy, but subtly modified it to the advantage of one or another category – in this way the bastards had gained a certain advancement, unjustifiably so in the opinion of Saint-Simon. At all events, in this text, Madame is not excessively tormented by the promotion of the illegitimate children since they are, in any case, far outranked by the princes of the blood who, in turn, are legally and in terms of advantages preceded by the Palatine's children. On the other hand, she has never come to terms with the fact that her granddaughter (for whom, in reality, she has little affection), is preceded by the married princesses of the blood. Here it is her own blood and above all that of her husband, the blood of Monsieur and Madame, which is depreciated as a consequence of the scheming of Madame la Duchesse. The latter, bastard daughter of Louis XV, married to a prince of the blood, is an influential participant in the cabal of the King's son: her influence is sufficient for her to have been able to inflict this insult on the Palatine's granddaughter. One may thus see to what extent the cabal interferes with rank.

Other texts give details of the postures (whether one is seated or not, and in what circumstances) by which rank is defined:

I see that you take my son for a prince of the blood. But he is not one. His rank is that of grandson of France: he is superior to the princes of the blood and has more privileges. Grandsons of France greet queens, sit in their presence and ride in their carriage; princes of the blood may do none of this. The domestics of grandsons of France benefit from certain exemptions and serve in turns. The grandsons of France have a first squire, a first almoner and a first major-domo. Princes of the blood enjoy no such privileges, and neither are they served by additional body-guards as is my son, nor by the Swiss Guard . . . (March 27, 1707)

Like Saint-Simon, although at times with more clarity and less passion, Madame tackles the problem of seats (armchairs, backed chairs, stools), which give material expression to the hierarchy's descent.

The King did not wish difficulties arising from court etiquette to be solved by half-way solutions. The Duc de Lorraine laid claim to an armchair before Monsieur and before me, since the Emperor conceded him one. The King replied that the Emperor had his own court etiquette, and that he, the King, had another: thus, for example, whereas the Emperor granted armchairs to the cardinals, they were not permitted to sit before the King . . . Monsieur was quite willing to give the Duke a backed chair and the King agreed, but the Duke claims to be treated as an Elector, and this was something the King did not wish to admit. Monsieur had proposed that we do as the King of England, who denies our pretension to be given a chair while we, for our part, sustain it, this being why he only ever sits on a stool in our presence . . . But the King would not hear of such things and, so as to avoid transforming an affront to the Duke into a diplomatic catastrophe, we decided against the journey to Bar . . . (October 1, 1691)

In this case, the "gradation" of the signs of distinction that Monsieur, the King's brother, would like to confer (but which are contested by the Duc de Lorraine) are easily read: armchairs for the Electors (they owe their elevation to the fact that it is their vote which determines the election of the Emperor); a simple backed chair for the Duc de Lorraine (who would like something more); and, thereafter, stools for the dignitaries who follow.

In the presence of Louis XIV and in certain circumstances, only the Queen of France *in partibus* – otherwise known as La Maintenon – has, scandalously, the right to an armchair; this abuse is disguised by reasons of health. None of the *vulgum pecus*, including the Dauphin, enjoy this honor.

I asked Monsieur why, in the lady's apartment [La Maintenon], I had only been offered a stool. He replied that nobody was allowed an armchair, not even the Dauphin or the Duc de Bourgogne, because the King came there very frequently. It is a good ruse, not to allow anyone to have an armchair, except for this lady, who is permitted to have one on the grounds of poor health! In this manner, she is neither said to be queen, nor not to be, and yet is ranked as such . . . (July 19, 1699)

The use of these material signs continues much lower on down the social scale, and as far as the dukes and peers; in the "bonnet affair" (which now appears ridiculous), they are mobilized against the "presidents with mortars" (Boislisle XXVI, 1).[4] Here, as elsewhere, it is not merely a question of difference between two social groups, but of an overall conception of the articulated and holistic sum of relations between men and between groups. Perched at the summit of society, Madame takes in the descending ranks in a single glance: children and grandchildren of France, princes of the blood, legitimized bastards, dukes and peers, etc. The organizers of processions in Toulouse were doing as much in distinguishing the gradations of the six categories; these were, in order, the Robe, the liberal professions, the commercial body, the bourgeoisie or persons of private means, arts and trades (a total of eighty-one hierarchically classed trades), and lastly, the "unincorporated" citizens. The social exaltation of the citizen of Toulouse is less than Madame is, but his classificatory spirit is of the same kind. This does not mean that the hierarchy of estates alone is what counts. It may be corroded by the money or power of one person or another. Likewise at court, despite the rigidity of rank, a King's valet may frequent the company of princes and snub that of dukes.

The question is not only one of establishing hierarchies or interdependence or encapsulation. The almost individual character of such a classification must be stressed. In this way the hierarchical spirit envelopes not only molecules, but atoms. The categories described by Madame – sons, then grandsons of France – can possibly each be reduced to one person in certain cases without mentioning the King himself; this does not prevent their comparison, or alter their gradation in terms of precedence with respect to the larger groups situated beneath them, such as princes of the blood and, *a fortiori*, the dukes and peers. And amongst these latter too, hierarchization means individualization. "Monsieur de Luxembourg, proud of his success and the general applause which greeted his victories, believed his position sufficiently strong to entitle him to second rank, rather than his former position in the eighteenth rank of seniority among the peers . . . He thus attacked the precedence of Uzès, Elbeuf, Rohan," etc. (the list which follows gives, in order, the seventeen first lay peerages).

The craze for classifying, for "taxonomizing," has something zoological about it. Take those famous portraits (by Saint-Simon) of courtiers where the personage under consideration, male or female, is reduced to the essence of an ox or an elephant, a cow, a large parrot, a boar, frog, lobster, snake, bird of prey, vulture, cat, monkey, lion, ferret, dog, horse, rat, eel . . . (Van der Cruysse 1971: 213ff.). From time to time, Madame resorts to comparisons of the same kind (describing herself as a bear, cat, or monkey). She is hostile to the Cartesian theory of animal-machines, yet this is normal given that Descartes is, in his own way, democratic in admitting that common sense is the best shared thing in the world. According to him, each man is his neighbor's equal, in so far as each constitutes a *cogito*. Descartes refuses a fragmented vision of humanity as a species; this refusal, for the philosopher of the *cogito*, is accompanied by a radical devaluation of zoology. He considers the animal as nothing more than a machine, the exterior or envelope of which is the equivalent of the casing of a watch or clock. On the other hand, the use of symbolic taxonomy for zoologizing man, in the style of Saint-Simon or Madame, constitutes an admission that in the end the human being may be put on the same lowly plane as the animal and, simultaneously, presupposes that our "inferior brothers" may, each according to its type, be matched with men. Within this perspective there is no question of comparing beasts with vulgar machines. As regards animal-machines, Madame remains resolutely hostile to

Descartes, the father of these weird concepts; she prefers Leibnitz who, in the same manner as, later, Rétif de la Bretonne and Konrad Lorenz, refuses the degradation of animal status to simplistic mechanics.

In much the same fashion as the Little Duke, Madame hierarchizes, taxonomizes and zoologizes. And, in so doing (again, like Saint-Simon), she sacralizes. For thoroughly hierarchic ideology presupposes the opposition of the sacred and the profane. Anthropologists who study India have made this point; it remains equally valid for our aristocratic hierarchies of the West. In Madame's case, the linking of the hierarchic and the sacred is all the more remarkable when one considers that her religion is otherwise enlightened, tolerant, skeptical, and already a prey to the "crisis of the European conscience" (as Hazard [1935: 25] was to call it). At an advanced age, Madame (who has, as she puts it, her "little religion all to herself") comes close to a sympathetic devotion which is from the heart, and which has nothing rigid or formal about it. But the moment one touches upon hierarchized social structures, madame's religion becomes ritualized and aspires to the ceremonial:

> It should come as no surprise to you that here, at mass, distinctions of rank are made. Thus, only the granddaughters of France have the right to a chaplain who gives the responses during mass and holds a candle from the *Sanctus* of the *Preface* until the *Domine non sum dignus*. Princesses of the blood are entitled neither to candles nor to individual chaplains and, during mass, responses must be given by their pages. At the end of mass, the priest presents the Corporal cloth to be kissed; this goes no further than the children of France. The same applies to the chalice in which the wine and water are offered; only *we* are entitled to drink from this chalice, which does not go as far as the princes of the blood. Hence you can see that ceremonial is here every bit as much in evidence as devotion. In this country there is, in all things spiritual, a concern for the temporal: if something desirable is seen as not being in accordance with the wishes of the good Lord, it may be redeemed by a temporal aspect: thus, as you can see, all is not lost . . . (April 3, 1710)

The scenario is evident: the children of France may kiss the Corporal. This cloth, on which the host is offered, is the supreme symbol of purity (we shall return to the problem of the pure and the impure with regard to hierarchy). Granddaughters of France have a chaplain who is entitled to give the responses during mass; they are thus allowed to participate directly, through the direct – albeit minimal – action of a certain "clergy" of petty clerics. Lastly, the princesses of the blood "make their pages give the responses in mass"; the page plays the role of a simple choirboy, essentially a lay person. The princesses are hereby at a slightly greater distance from the sacred than are the above-mentioned female personages of higher rank, but they retain a certain contact with it. (Note that in this passage noblewomen participate in the mysteries of religion through the intermediary of their higher domestics, clerics or pages; these are a part of their family and even their person – there is the famous saying of two sisters of the highest nobility, who wished to do penance for their sins. "Sisters, let us make our servants fast.")

As soon as she touches on the question of rank, then, Madame envisages a religion which differs from that of her usual devotions, sincere at times and skeptical at others. Now it is a matter of a more archaic religiosity. The spiritual informs the temporal, and vice versa, which runs counter to the wishes of more modern seventeenth-century authors, who abstract religious fervor from its social context.

A detailed account of the hierarchy of ranks in connection with the Eucharist is to be found in another text, this time from Saint-Simon.

> There I learned of a new undertaking of the princes of the blood, who, kept in an impotent and discredited state by the King, took immeasurable advantage of his desire to advance his bastards, whom he had assimilated to their ranks, in order to acquire new privileges which were allowed them on condition that they be shared. The superiority and differences of rank so clearly marked above them by the grandsons of France had always been hard to bear. One such distinction arose at the King's Communion where, after the elevation of the mass, a credence[5] was pushed to the base of the altar, to the place where the priest begins his Communion. It was covered with a sheet, and then with a large cloth which fell to the ground in front of and behind the table. At the Pater the almoner of the day rose and whispered in the King's ear the names of all dukes in the chapel at that moment. The King named two of them, these always being the two most senior; the almoner immediately presented himself before the said dukes and bowed to them. During the priest's Communion, the King rose and went to kneel behind the credence, with neither carpet nor mat, and took hold of the cloth; at the same time the two dukes, who had been informed of their role, alone with the captain of the guard on duty, rose from their places and followed the King, the senior on the right, the other on the left, and each took hold of a corner of the cloth, kneeling at the same time as the King and nearby him. The two almoners of the day likewise took the other two corners of the cloth, on the altar side, all four of them on their knees as was the captain of the guard, alone behind the King. Having received Communion, followed by ablution several moments later, the King remained briefly in the same position, then returned to take up his own place, followed by the captain of the guard and the two dukes, who likewise resumed theirs. Were there but one son of France present at the King's Communion, he would hold the right corner of the cloth, with no one on the left, and when Monsieur le Duc d'Orléans attended and no sons of France were present, the same rule applied: no prince of the blood could serve alongside him. If, however, there was only one prince of the blood present, one duke rather than two was designated in the usual manner, and served on the left, the prince of the blood serving on the right. The King named the dukes to show that he was master of the choice amongst them, and was not constrained by questions of seniority; yet on no occasion did he prefer any other than those who had served the longest. I remember walking ahead of him on the way to the chapel for Communion and the King, espying the Duc de la Force, addressed the Maréchal de Noailles in hushed tones; a moment later, the Maréchal came and asked me who, out of Monsieur de la Force and myself, had served the longest. He himself was obviously not certain, and the King wanted to know, to avoid making an error. The princes of the blood, wounded by the distinction accorded to Monsieur le Duc d'Orléans, from which in fact they had little suffered prior to his going to Spain, wished to avenge themselves by

usurping this privilege from the dukes. They carried out their affair in secrecy and, at Assumption of this year, Monsieur le Duc served alone at the King's Communion, no other dukes being called upon. I informed Forges of the matter. I knew that it had met with great surprise, and that the Duc de la Force (who should have served) and the Maréchal de Boufflers were both in chapel at the time. I wrote to the latter, not long ago, that such an occasion was without precedent; that I myself had served alongside the princes of the blood and even Monsieur le Duc, that this was easily verified on the register kept by Desgranges, Master of Ceremonies; and what I believed it was necessary to do in order not to suffer another loss. Upon visiting the registry, what I had said was found to be written and approved, and many other examples besides; the weak and wretched dukes dared not take matters any further.[6]

Foreseeing this, I had written, at the same time, to Monsieur le Duc d'Orléans, in Spain, recounting all that I judged most apt to irk him. As regards the maintaining of his distinction over the princes of the blood, I warned him not to put up with their usurpation of the dukes' privileges, since this would, in so far as possible, render them equal to his own person. Upon his return, I ensured that he mention the matter to the King; the latter, however, declined to comment. Monsieur le Duc said that there had been no question of interests, but, upon the insistence of Monsieur le Duc d'Orléans (timid as he was with the King), the King replied that it was the dukes themselves who had not come forward. Yet how could such a thing have come about without their being forewarned, and how had this affair been deemed by the King himself? In fact, nothing more was made of it, and so things remained. (Boislisle XV, 237ff.)

This passage presents the full scope of the problem of the predominance of sacred values, compared with those of a royal, princely, or ducal king (the almoners are ranked above the princes and the dukes; the King kneels on the bare ground). The side immediately facing the altar is preeminent over the front side of the nave, and the right side has precedence over the left. The degraded laity, facing the body of Christ, descends from the King down to the sons of France, followed by the grandsons of France (the Duc d'Orléans), the princes of the blood (ever trying to improve their position), and the dukes and peers, ranked according to seniority. Thus, yet again, the taxonomy reaches as far as individuals or their respective lineages.

The ceremony which accompanied the taking of the veil also furnishes a number of views of the hierarchization of physical stance: those placed beyond the sacred space demarcated by Madame's prayer-mat are on their knees, whereas those in front of the ecclesiastical princes are "flat on their stomachs" in these (admittedly exceptional) circumstances. In the following case it is Madame's granddaughter, daughter of the Regent, who is taking the veil; although Madame, Protestant at heart, detests the convent and the nuns, she describes this episode most conscientiously since, yet again, it is a question of the sacralization of rank:

We reached Chelles at half past nine; my grandson, the Duc de Chartres, had already arrived. We went to the church shortly after ten o'clock. The abbess' prie-dieu was in the nun's choir, and was of purple velvet completely covered with gold-embroidered fleur-de-lis. My prie-dieu was against the balustrade of the altar, and my son was at my side; his daughter was behind my chair, for princesses of the blood do not have the right to kneel on my prayer-mat, this right being reserved for the grandchildren of France, such as my son and my daughter . . . The Cardinal de Noailles conducted the ceremony . . . After he had read the epistle, the Master of Ceremonies went to the nuns' choir and sought out the abbess. She came forth, made a deep reverence before the altar, and another before me, and ascended the steps and knelt before the cardinal, who was seated in front of the altar in a big armchair. The abbess' confession of faith was ceremonially presented to her; this she read, then prostrated herself on the last step of the altar.

(September 17, 1719)

Once again, the hierarchy between children of France, grandchildren of France and princes of the blood is manifest down to their very kneeling in church. At the same time, these different groups are organized around Madame's prayer-mat according to the hierarchies of the sacred. She outranks all three groups because the young and newly received abbess, princess of the blood by birth, is prostrate before her God, momentarily represented by a cardinal.[7]

Lastly, the sacred is present not only at the summit of the court's hierarchy, but also throughout the entire length of the descending ranks which demonstrate to the whole world the further hierarchy's links. The Order of the Saint-Esprit, reserved for the nobility, illustrates a particular devotion to the third person of the Trinity and, at the same time, an acute sense of the descending hierarchy; each member of the Order occupies a certain position in the processions (Boislisle I, 61, etc.). Saint-Simon and others have written at great length about this Order and its ecclesiastical knights; such information seems little comprehensible in today's egalitarian and disenchanted world. The fact that this Order of the Saint-Esprit exalts, simultaneously, the numeration of men and of the sacred remains an inexhaustible subject. The Order celebrates three great festivals annually, two of which are consecrated to purification (Candlemas) and spiritual infusion (Pentecost); in the course of these ceremonies, at least until 1661, the Knights of the Saint-Esprit took Communion in full dress (Boislisle XXVIII, 273). The knight, upon receiving a collar of the Order, uttered those same words which were addressed to Christ by a sick man, about to be miraculously healed: *Domine, non sum dignus* (Lord, I am not worthy, but one word and my body shall be healed) (from article XXV of the primitive statutes of 1578: Boislisle XVI, 57 n. 4).

More generally speaking, the sacred permeates a large part of the French population which, even as late as 1700, remains constrained within the cells of corporate society. The Court is bathed in the sacred: mass, vespers, and the Holy Sacrament succeed one another in the Versailles Chapel, which is raised one floor above the rest of the palace. Does the Court form a single body, or rather an ill-assembled aggregation of bodies, at times disunited? Whatever the answer, at the much lower level of craftsmen or villagers one can see the guilds and somewhat hastily labeled corporations assuming succession of the brotherhoods; these organizations are, in turn, grouped around the cult of a patron saint (who may still be the Holy Spirit, for urban and village communities of south-east France, Saint Crispin for the shoemakers, Saint Eligius for the blacksmiths, etc.). Even today's unions, distant heirs of corporationism, are not conceived free of an ideology which constitutes a surrogate for religion (Debray 1981).

In its various summits and essential links, the hierarchy of the court is bound to the gradations of celestial origin. And yet the opposition of the sacred (the supreme value) and the profane (the other extreme of the same axis) does not imply religious intolerance: in fact, to set the sacred at the summit of a hierarchy is to allow that other, more lowly segments of society may legitimately escape the aura of this supreme value. The authors who most acutely felt those court hierarchies to which they clung with every fiber – and I am referring to Saint-Simon and the Palatine – were manifestly open to religious pluralism. *A priori* they were not at all hostile to Jansenism (except when Jansenism seemed to manifest a political opposition), despite its persecution by those in power, and acts of pure and simple intolerance such as the Revocation of the Edict of Nantes were total anathema to them. The Huguenot exclusion (1685) coincides with the formation of a modern national State in France, its egalitarian tendencies already evident: they demarcate or separate full members of the State or nation from those scapegoats whose exclusion constitutes a further affirmation of national cohesion. Hence in France and in the British Isles the years 1680–1700 are marked by a wave of persecutions: of the Huguenots south of the Channel, and of the papists or Irish in the north. Again, the fact that a *homo hierarchicus* like Saint-Simon and a *domina hierarchica* like Madame (admittedly of Protestant origin) are not remotely connected with these exclusionist waves or tempests once more bears witness to the nationalist modernity of such movements. Society based upon rank, such as that incarnated by the Little duke or the Palatine, could hardly object to counting certain Protestant lords amongst its gradations, irrespective of their importance.

Similarly, Madame and Saint-Simon display tolerance towards Jews, antisemitism appearing in the letters of the one and the Memoirs of the other to an infinitesimal degree, if it is there at all.

Hierarchy is animated by the sacred like iron filings by a magnet: on the other hand, it loathes the impure. Thus, the opposition *sacred*/profane is corroborated by the contrast between pure and *impure* (in each case, the strongest term is italicised, *sacred* as the height of the scale in the first instance, and *impure* constituting the lowest point in the second case).

Louis Dumont (1976) has dealt with this problem at length, with regard to the castes of India: the purest castes, like the brahmins, must refrain from physical contact with those of lower castes, and purify themselves of bodily secretions (excrement, menstrual pollution, etc.). Certain contacts are to be avoided, and those belonging to such groups bathe frequently. Even the pipe may not circulate from the lower castes towards the higher orders, because its very mediation would imply contact between the lips of its successive users, pure and impure.

In an extremely hierarchized society like the Court (which schematizes the cumulative degrees of contempt that characterize the West) the situation is different. First, there are no castes, strictly speaking, either in the near or distant entourage of Louis XIV. There are simply the different ranks. Second, ideas relating to bodily cleanliness are as demanding for the Sun King's eminent contemporaries as for the brahmins, but they concern regions of the body which are not the same.

Whereas the brahmin cleanses the exterior of his body (in other words his skin), French high society of 1700 attributes the greatest importance to purification of the interior, by purges and blood-letting. Purgatives and the lance. Taboos on excrement exist but they are weak, much more so than nowadays. Saint-Simon's Memoirs devote pages to *chaises percées* and successful purgations, and bleeding seems to be necessary for vital balance; our Little Duke is the first to assert the cardinal value of such techniques. The higher one's social position, the more one is bled and purged – the young Louis XIII, eventually to die of such methods, was all too well aware of this distinction. Saint-Simon raises not the slightest doubt about the legitimacy of blood-letting or intestinal cleansing. He never directs any blame at the surgeon's corporation (even when bleeding results in an abcess), and has the greatest esteem for Mareschal, first surgeon to the King. Madame has some reservations; nevertheless, for better or for worse, she accepts the regimes of purgation and bleeding to which she, along with all the highest dignitaries of the realm, is subjected although grumbling. She even ends up somewhat revived by these procedures of inner cleansing of the bodily systems.

The obsession with purity thus occurs in a different context than it does for the Indians. While there is little concern for excremental and menstrual contact, it is the stains of illegitimate birth and, more generally, sexual distortions which are incriminated. The Court, together with the entire Western world, is prey to the "monarchy of sex" (as Michel Foucault calls it).

First there is the problem of the royal bastards. They bring the greatest impurity (bastardy) into the heart of what should be the site of the system's ultimate purity (the royal person and family). These illegitimate offspring constitute a form of absolute scandal, and it is perfectly normal for Saint-Simon to attack them with such extraordinary vigor (as does Madame, albeit incidentally). Such attacks appear superfluous to the twentieth-century reader, in so far as his egalitarian viewpoint renders him no longer able to comprehend the demands of a hierarchy-based system. In fact, the impurity of being born of an adulterous or illegitimate union well and truly contradicts the values of the sacred; from 1597 onwards, the statutes of the Order of the Saint-Esprit proclaim that "no bastard may be admitted to the Order, save recognized and legitimized bastards of the King" (Boislisle XV, 43 and n. 1); an exception which, needless to say, displeases Saint-Simon. Being partially excluded from the Order, the bastards are dispossessed of the role of bearers of the cloth at the King's Communion. Saint-Simon's language is violent on the subject of royal bastards, male and female, to whom Louis XIV "shamelessly" married princes and princesses of the blood, and even a grandson of France. The Little Duke's vehement writings emphasize the fact that the very notion of purity seems to be compromised by such intrusions of illegitimacy. There are dozens of pages in the Memoirs which could be quoted. I shall limit myself to one (amongst many). Its opening is thunderous, and denounces du Maine, the very incarnation of the royal bastard (Boislisle XXVIII, 310).

The monstrous marriage of Monsieur le Duc de Chartres, later d'Orléans, and Regent, that of Monsieur le Duc, those of the daughters of these marriages with Monsieur le Duc de Berry and with the Prince de Conti, have brought about what the King has viewed with his own eyes, and viewed with complacence: that is, apart from his unique successor and the Spanish branch (although in fact the Renunciations have excluded the latter from the line of succession), there is no longer one single person in the royal household, male or female, who is not the direct issue of the King's liaison with Madame de Montespan. This woman is either mother or grandmother of all – even if the Duchesse du Maine is not one of her own offspring, she has married the son of the King and Madame de Montespan. The only daughter of the King and Madame de la Vallière has married the elder of the two princes de Conti and borne no children, but it was no fault of the King's that this single branch of princes of the blood escaped bastardy until it, too, was finally tainted thereby, in the second generation. Let us not forget that it was the Prince of Orange's refusal of this princess – refusal which no amount of respect, desire, care and the most lengthy submission have managed to efface from the King's heart – that has made this famous prince "in spite of himself" the enemy of the King and of France. This hatred, the source and fatal cause of all these leagues and wars, whose weight has so nearly brought about the surrender of our King, is the fruit of that same bastardy which may rightly be called the fruit of destruction.

This mingling of the purest blood of our kings, and one may frankly say of the whole

Figure 3.1 The sacred/impure opposition

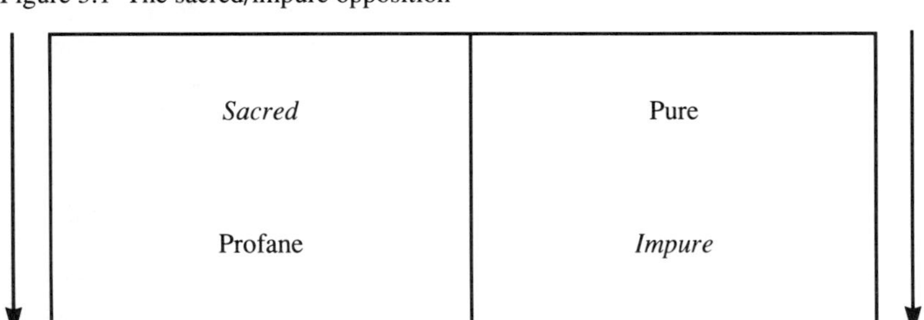

universe, with the corrupt filth of twofold adultery, has thus been the King's constant task throughout his entire life. He has had the horrible satisfaction of fusing them together and hence taking to its limits a mixture unparalleled in all history, after having been the first of all men, of all nations to have dragged out of nothing and brought into existence the fruits of twofold adultery, at first repulsive to the whole world, civilized and barbarian, but which he has since managed to have accepted as normal. Whereas the recognition and protection of the bastards assures the path of fortune, that of the princes of the blood, beginning with Monsieur, has ever constituted an insurmountable obstacle. Such have been the fruits of limitless pride, and this is why the King looks upon the bastards and the princes of his blood with such different eyes: the children issue to the throne by legitimate generation, and those born of his liaisons. The former he considered as children of State and Crown, thereby great in their own right. The others he cherished as children of his own person; whatever the laws, they could know no independent existence, and could never amount to more than works of his own power and his own hands. Pride and tenderness united in their favor, and while the vain pleasure of creation grew ceaselessly, it was ever stabbed by jealousy of the natural independence enjoyed by the others, by those whose stature did not require the King's intervention. Irked by the fact that he dared not equal nature, he at least contrived to bring his bastards nearer the princes of the blood in all that he first gave them, by way of establishments and rank. He then tried to confuse them by means of monstrous unthinkable marriages, multiplied so as to make of them one and the same family. The only son of his only brother was at last sacrificed too, and this with the most overt violence. Then, having become more daring through his redoubled efforts and triumphs, the King established complete equality between his bastards and the princes of the blood. Finally, preparing for death, he forgot all constraint and gave them the name and right to succession to the Crown, as though it lay within his powers to make of men that which, by birth, they are not. This was not all. All his care and last measures for after his death were but in their favor. Artfully alienated from his nephew, and carefully maintained in this attitude by the Duc du Maine and Madame de Maintenon, he suffered the yoke he had allowed them to impose upon him; he drank from the chalice which he himself had prepared. Then his outbursts of resistance and bitter regrets were to be seen; he was helpless before their extortion. His successor was entirely sacrificed, and, inasmuch as they went together, his kingdom. Everything which was named in anticipation for the education of the future King had as its sole motive the interest of the bastards; nothing else and nothing less. The Duc du Maine was placed in charge, and under him the Maréchal de Villeroy, possibly the most inept man for this task in the whole of France . . . In this matter, if in nothing else, the King's memory cannot be cleansed, either before God or before man. This was the last abyss into which he was led by vanity and weakness, a more than obscure woman, and twofold adulterers to whom he abandoned himself, of whom he made his tyrants, after having tyrannized them along with so many others; they abused him without scruple and without reserve. Onto this path he was also led by that detestable character, his confessor, Father Tellier. Such was the repentance, penitence, of public reparation for the flagrant twofold adultery which had been so flagrant, so long, and so scandalous in the eyes of all Europe, and such were the last sentiments of a soul so highly sinful, preparing himself to appear before God, and, moreover, responsible for a reign which had lasted fifty-six years, and which had been marked by pride, luxury, edifices, profusion of all kinds and continual war. The vanity which had been source and nourishment for all this had caused incalculable bloodshed, consumed innumerable millions, both within and without the country; it had incensed all Europe, confusing and annihilating all order, all rules the most ancient and sacred laws of the State, reducing the kingdom to irremediable wretchedness, and bringing it so imminently near to total loss that it was saved only by a miracle of the Almighty.

This text contains everything – even if it verges on madness! The raising of the bastards was monstrous, it perpetrated the corruption of double adultery which was spoiling the royal blood and tainting the memory of the Monarch. It meant war and wretchedness.

It also constituted a violation of the most sublime manifestation of the sacred (in raising his bastards to the height of his legitimate successors, the King equaled himself to God; in the manner of the Creator, he wished to raise them out of nothingness). Finally, it was the stupidity and infamy of a woman of little worth (la Maintenon) rendered delirious by her social ascension. Crowning this vast edifice was, of course, impurity, "this mingling of the purest blood of our kings, and . . . of the whole universe, with the corrupt filth of twofold adultery."

Hierarchy thus opposed the sacred and the profane, but also blood and filth; this filth was scarcely better than the dregs of society, whence come several persons of low extraction such as Father Tellier, Cardinal Dubois and our author's other *bêtes noires*.

If bastardy constitutes a distortion of procreation, sodomy, according to the thought of the time, was a form of sexual expediency; on several occasions both Saint-Simon and Madame thus link this vice to illegitimacy on the register of the impure. Hence, it is as though homosexuality were the ultimate source of aggravation and impurity. D'Effiat and the Chevalier de Lorraine, homosexuals and lovers of Monsieur, the King's brother, negotiate on la Maintenon's behalf the marriage to be brought about between Philippe d'Orléans, son of their friend Monsieur, and Mademoiselle de Blois, bastard daughter of Louis XIV. Here, the links between sodomy and twofold adultery are outwardly condemned by the Palatine and the Little Duke. These criminal bonds increase yet further, if that can be said, the maximum impurity engendered by bastardy and the dangers which stem from it. We might add that d'Effiat and the Chevalier de Lorraine are considered (probably wrongly) by Madame and Saint-Simon to have been the poisoners of Henrietta of England. Sodomy, poison, and then support for bastardy: these are all associated on the register of the impure. In much the same way, Cardinal de Bouillon, one of Saint-Simon's black sheep, may be seen to incarnate all that can possibly displease; he is homosexual; he is linked with the Jesuits (who, according to Saint-Simon, are masters of cunning); and lastly, and very logically, he seeks to pervert rank and hierarchy (did he not dare retain his red calotte on his head in the presence of the Pope, thus scandalizing the other cardinals?).

Another major impurity is to do with sexuality. It concerns venereal disease and syphilis. Bastardy, sodomy, and the pox combine to blacken absolutely the personality of Vendôme who, apart from Saint-Simon's imprecations, has gone down in history as being a very capable soldier. Vendôme descends by legitimate male lineage from one of Henri IV's bastards. Saint-Simon, ever punctilious as regards purity of the blood (especially with regard to those whom he detests) can find no excuse for this blot on the escutcheon. Furthermore, Vendôme is a sodomite and makes no secret of the fact. Lastly, syphilitic to the point of losing his nose, he has managed to crown it all by marrying a princess of the blood, Mademoiselle d'Enghien. This marriage was arranged under the auspices of that epitome of all bastards, the Duc du Maine (Boislisle XIX, 111). Hence the royal heredity of the daughter of Condé is sullied three or four times over by the multiple and cumulative taints of her husband.[8]

An opposition is evident in this passage: on the one hand feature the high values of physical purification brought about by purging, with the happy consequences they entail; on the other hand, there is the ultra-pejorative nature of sodomy. As regards the *Chateau de derrière* (as Madame calls it), it is recommended that the flood gates be opened; but it is ill viewed that they be taken by assault. Exit is praised and viewed as a benediction: whereas entry – forced or fully conceded – is an abomination.

The writings of Saint-Simon and, to a lesser extent, those of Madame, contain a geography of the impure and the pure. With sodomy, we are in the South: a Greek and Italian vice. The Letters and the Memoirs constantly bring us back to these peninsulas. In so far as bastardy is concerned, Spain scarcely does better: "Almost all houses of importance and distinction in Spain are bastard houses, often more than once over; this bastardy envelopes almost all the greatest and noblest lords in the country" (Boislisle IX, 160–250). The author of the Memoirs explains this vexatious adulteration by former influence of the Moors, with their "commerce of mixing" with the Iberians, "almost up to the reign of the Catholic Kings" (ibid.). In other words, the impure in this case connotes both the lack of the sacred and the frailty of Christian holiness, so long under the yoke of Islam.

On the other hand, poles of purity as regards respect for rank and birth are situated in the North, let us say in the Germanic areas and to a lesser extent in England. The German nobility abhor bastards and detest misalliances. Madame is the living example of this attitude, which one finds underlined time and time again, in both the Memoirs and the Letters. Upon learning of Louis XIV's decision to marry her son to his royal bastard daughter, she slaps the unfortunate young man, and her acknowledgment of the Monarch's reverence on this occasion is so brisk that "the latter, upon straightening up, saw no more than her back, and [Madame] already one step nearer the door."

Saint-Simon would like to annex England to this geography of countries vaunting hierarchy; he makes a passing reference to the fact that Berwick "very English could not suffer inversion of rank" (Boislisle XX, 270). The matter is one of topical interest since Berwick is hostile to d'Antin's attempts to obtain as Seigneur d'Eperon the dignity of duke and peer, to which he is not entitled. D'Antin, legitimate son of la Montespan, is using "his mother's condemnable fecundity" (Boislisle XX, 261) to win the King's support for his case. Berwick, bastard of an English King, thus wishes differences in rank to be respected; and is opposed to the legitimate son of a mother of bastards, even if she is mistress of a King of France. We shall not dwell upon this paradox, but state simply that Saint-Simon harbored certain illusions concerning England's case; in reality, the British nobility was more open towards the upward mobility of commoners than was the French (and perhaps the German?). The fact however remains that, in this geography of the fantastic, sodomite Italy and Greece and bastardized Spain are each contrasted with "good Germany" which respects hierarchy established by birth. France, between the North and the South, is suspended between these areas and hesitates between the two models – the impure Mediterranean and pure Germanic worlds.

Leaving the sacred and the impure, let us revert to the central notion of hierarchy: this notion conditions our authors' views of court conflicts. Several modes of classification may be discerned:

- A Royal couple, children of France, grandchildren of France (these three levels define the crest of the hierarchy).
- B Royal family, properly speaking, versus princes of the blood (the hierarchy is structured as above, but at a slightly inferior level).
- C Legitimate offspring versus bastards, or legitimate union versus royal adultery (this third axis opposes the pure and the impure, manifesting the extreme poles of the hierarchy).

This first scale, "A," is by far the most useful for analysis of court conflicts and cabals. Madame and Saint-Simon saw little of each other and did not converse on this matter (in social terms, the Little Duke was small-fry for the great lady and she situated him, along with his peers, a good hundred feet lower than the Palatine counts); furthermore, their respective texts remained unpublished until the nineteenth century. Hence one may consider that the unanimity so clearly demonstrated by this analysis is witness to the Court's vision of itself (at least in those passages which have the strongest tendency towards some kind of sociological self-reflection).

Our first quotation comes from a letter written by Madame to the Duchess of Hanover, dated September 28, 1709 (Jaeglé collection, vol. II, pp. 101–2):

> The entire court is full of intrigues: there are those wishing to obtain the favor of the powerful lady, others seeking that of Monsieur le Dauphin, and others again seeking that of the Duc de Bourgogne, since he and his father are on bad terms. The son despises

the father, is ambitious and wants to govern; the Dauphin is completely dominated by his bastard sister, Madame la Duchesse. The Princesse de Conti has become the Duchesse's ally, so as not to lose her hold over him completely. All are against my son: they are afraid that he is out of favor with the King, and that he will marry his elder daughter to the Duc de Berry. The Duchess wants the Duc de Berry for her own daughter, which is why she monopolizes him. However, the Duchesse de Bourgogne, who would also like to govern the Dauphin as well as the King, is jealous of Madame la Duchesse, and has thus made a pact of friendship with our Madame d'Orléans so as to foil the other's plans. This network of intrigues makes for an amusing comedy; I think that, as the song goes, "If one didn't die of hunger, one should die of laughter . . . " In this world the Old Woman pits one against the other, all the better to govern.

(Boislisle XVIII, n. 4)

Further on, she says:

I shall send you an account, through Monsieur Schoulembourg, of the state in which the court is presently to be found. The King is more than ever charmed by his aged beauty: everything passes through his hands, and everything goes askew, like the old lady's waist. She counts on gathering her profits, on gaining money from all things and on teaching this craft to the Duchesse de Bourgogne; she knows all the State secrets and passes them on to the Duchesse, so that nothing remains hidden. Monsieur le Dauphin is still in love with la Chouin, and has perhaps married her. That Chouin is a cunning one; she refuses to appear at court, because if she came she would have to submit to the authority of the Dauphin's mother-in-law, and so stays well away. She sees only the Duchesse de Bourgogne and her favorites, and Madame la Duchesse, who is Monsieur le Dauphin's favorite. She accords them a singular honor when she permits them to dine with her: it is a question of what is here known as the *parvulo*. The most secretive and private of matters. La Chouin has her creatures everywhere; the Maréchal d'Huxelles, Albergotti, and the Counsellor of State Monsieur Bignon are also her counsellors. Monsieur le Duc de Bourgogne is leader of the devout faction: Monsieur le Duc de Beauvilliers, the Duc de Chevreuse, and Monsieur d'O. This cabal does not always look approvingly upon the other two, and often opposes them in counsel; the entire court is divided up amongst the three of them . . . (December 23, 1710)

In these two texts Madame thus refers to three cabals, lined up along the multigenerational axis of royalty. These are: the *father*, Louis XIV, and his wife la Maintenon; the *son*, the Grand Dauphin or Monseigneur and his mistress, if not secret spouse, la Chouin; and, lastly, the *grandson*, the Duc de Bourgogne. Historians consider the latter to have been the leader of an enlightened and liberal group (Fénelon, Beauvilliers, etc.), but Madame sees him chiefly as the head of a *camarilla* of the devout (in this, she is not entirely wrong).

A superb text by Saint-Simon (Boislisle XVIII, 5–19) also gives the pattern of the three cabals: the first – *father* – is conservative, lordly and military (that of la Maintenon); the second – *son* – is quite simply involved in intrigue (that of Monseigneur); the third – *grandson* – is intelligently reformative and ministerial, despite a smattering of Jesuit bigotry (that of the Duc de Bourgogne). For this text (corresponding to Madame's), one may refer to my particular study (Leroy Ladurie 1978, II: 275–99).

A quarter of a century earlier (letter from Madame, August 11, 1686), Louis XIV was not yet really old, and his succession was scarcely a matter of concern. The tripartite opposition between King/Grand Dauphin (born in 1661)/Petit Dauphin (born in 1682), or between Monarch/Son/Grandson, had not come into being, if only because of the youth of the latter two protagonists. Another device had to be found, to describe the "general plan of the court" in 1686. Madame drew upon the opposition existing between the immediate royal family and the granddaughters of France and princes and princesses of the blood. The close royal family extends from the cryptomonarchic couple, which, even at that time, was constituted by Louis XIV and la Maintenon; the granddaughters of France are the two daughters of Gaston d'Orléans, brother of Louis XIII (Mademoiselle la Grande Duchesse, or la Grande Mademoiselle, and Mademoiselle de Guise). The princes and princesses of the blood are the male and female descendants of the Conti and Condé-Bourbon lineages (Monsieur le Duc, Madame la Duchesse, the Princesse de Conti, Madame de Bourbon . . .). Hence the following description (August 11, 1686).

Truly, someone who had nothing to do with this court would laugh until they cried on seeing how things happen here. The King imagines himself to be devout because he no longer sleeps with young women, and all his fear of God consists of being obsessed by detail, of having spies in all corners who make wild and false accusations; of flattering his brother's favorites, and of generally being a torment to everybody. The old woman, la Maintenon, takes pleasure in making all members of the royal family quite odious to the King, and in ruling them; all, that is, except Monsieur (the King's brother), whom she flatters to the King . . . From what I have said, you can more or less see one general plan of the court as it is at the moment: neither good nor ill is spoken of Mademoiselle la Grande-Duchesse and Mademoiselle de Guise, who are looked upon as being nothing, which I myself would consider a blessing, and would willingly change places with them. Monsieur le Duc is *crawling on his stomach* before all that might be called favor, and people mock him, amongst other things, because of his banality. Madame de Maintenon plays off the Princesse de Conti against Madame de Bourbon as though they were sitting on scales, raising this one and lowering that one, favoring one and dismissing the other. At the moment, Madame de Bourbon is in favor and the Princesse de Conti is in disgrace, but things will change before long. The reason for the Princesse de Conti's disgrace these days is that spies have told the King that she and her cousin, the Duchesse de Choiseul, made fun of la Maintenon. Madame la Duchesse is no happier than the rest of us, for her husband is a real tyrant to her.

The import of this description is infinitely narrower than that of the trigenerational design (King/Monseigneur/Bourgogne) referred to in the preceding paragraphs. For according to this hypothesis, conflict between the three generations of the royal family – or, rather, between their partisans and all those implicated – depended upon wagers involving the delays and modalities of succession to the monarchy. These hushed conflicts thus took in the more extensive elements of power – ministers, chief bureaucrats, military leaders, prelates, Jesuits, etc. In 1686, however, true power apart from Louis XIV's was wielded by the mandarin-like dynasties of Letellier-Louvois, the family of the deceased Colbert, the Phélypeaux, etc. The quarrels which enveloped the Condé or Conti families are not as important as those opposing the Le Telliers and the Colberts. It is nonetheless interesting to note that Madame, faithful to her hierarchic vision of the world, utilizes from 1686 onwards the schematized summits of this hierarchy (the higher levels of the royal family) to decipher or decode the court system. The result of this analysis is defective, but it does not invalidate the basic principles of a perfectly curial mentality.

The years 1711 to 1715 saw the successive passing away of the Grand Dauphin, the Duc de Bourgogne, and Louis XIV. Thus, the trigenerational diagram is no longer relevant. As of this moment (after 1715), Madame will provide another schema of the court (which has, furthermore, been split up, through its transfer to Paris). Henceforth she will utilize the opposition of the pure and the impure (in other words, legitimate/bastard), to decode the conflicts of power. On the one hand there is the Little King, Louis XV, frail scion of the Duc de Bourgogne, and the Regent, Philippe d'Orléans, born legitimately of the "blood" of Louis XIII. Philippe, as Regent, defends the King, and leans on the financier Law (or *Las* as he was then called). On the other hand, at the lower end of the scale of impurity is the bastard Duc du Maine, raised by the old witch Maintenon, and supported by Parliament. This opposition is rendered all the more dramatic by the fact that Maine is brother-in-law to Philippe d'Orléans through the latter's wife, likewise a royal bastard. Here are Madame's views on the subject (July 14, 1718):

> It is all too true that Parliament has embarrassed my son . . . He told me that those Gentlemen interfered with affairs which did not concern them, and that as long as royal authority lay in his hands, he would keep it intact, and pass it on to the King upon his coming of age such as he had received it; he would thus not suffer anyone's touching it. Until now there had been nothing to fear; the people have not shifted and neither have the other Parliaments in the provinces. The brother of my son's wife [the Duc du Maine] and his wife are his worst enemies; these persons are turning everyone against him. Had he but listened to me, he would not be their brother-in-law, and would be able to act without having to fear tears. My son must find new means to pay the debts of the deceased King. This Las who is so hated is a very intelligent Englishman . . .

This opposition of legitimate/bastard, or Orléans/Maine, is too simplistic, failing to take into account the complex realities of 1718. Saint-Simon, now a direct participant in the power structure, paints a much more sophisticated picture of this period. On the other hand, during the years 1712–15 when the court was still at Versailles, but deprived through their premature deaths of the Grand Dauphin and the Duc de Bourgogne, Saint-Simon's explanatory model also drew upon the legitimate/bastard or Orléans/Maine opposition. It largely served as the key for his reflections on court and power over this three-year period. And for those of Madame, a little later.

Prior to the wave of egalitarianism beginning with the Age of Enlightenment, it is the court which forms the last *chronological* step of the society of rank. In the writings of Saint-Simon, Madame and several others, the court attains, if not a clear consciousness of its ideology, at least the perfect expression thereof. The spatial concentration of this ideology – which is confined to the limits of a large palace – renders the various gradations still more apparent. Hierarchies are more complete and behavior is more clearly symbolic, less aggressively violent. In this sense – but in this sense alone – curialization of the warriors (or, at times, of ex-warriors) represents, as Elias (1974) has remarked, a considerable step forward on the road leading to social rationalization. Seen from this angle, such curialization is in effect an aspect of modernization; in other respects, it remains archaic, for it expresses social hierarchies which are fully accepted and formed. These will lose their legitimacy, if not their existence, in the course of the egalitarian phases soon to follow. After 1715, 1750 or 1789 – from Saint-Simon to Rousseau.

A society of rank differs from a society of theoretically "pure" castes, where marriage may take place only between equals. Besides, the very structures of the caste system are neither genuinely nor systematically endogamous: female hypergamy is practiced; a family situated in an inferior group, rich in money but not in prestige, willingly offers daughter and dowry to a man of a superior group in order to acquire a certain prestige. Furthermore, this "offering" of daughters and, on the other hand, the warrior vocation of superior groups (leading to the extermination of a fairly large number of men of the elite) are factors which combine to produce a preponderance of women in higher social strata; men become a slight minority. The phenomenon of polygyny thus occurs, to the advantage of the most important persons.

There is no need to stress the polygyny of Louis XIV. It is simply an individual phenomenon, and does not, by any means, necessarily result in a typology. The wars of the Augsburg League and the Spanish Succession depopulate the court of its young and not-so-young males, who are killed, mutilated, imprisoned, or simply engaged in combat in Flanders, Germany, Italy, Spain. According to Saint-Simon, the court at times appears as nothing more than a great herd of sobbing women led by an ageing shepherd: they anxiously await news of the most recent combats, whilst the old King stoically suffers being informed by urgent missive of the successive defeats of his troops.

The important point here is feminine hypergamy, a means of climbing upwards over the barriers of rank, and a device assuring social and conjugal circulation in a society which, without such means, would seal off its hierarchical compartments.

Independent of Saint-Simon's and Madame's essentially qualitative intuitions, Labatut's statistical study (1972) provides the indispensable answer to this question. Forty-nine percent of the French dukes and peers of the seventeenth century marry women from their own ducal group (this being quite simply a question of isogamy); percentages for marriages of these dukes with daughters of non-ducal nobility (that of the Sword and the Robe) are respectively 37 per cent and 14 per cent. These two figures, and especially the latter, represent the proportion of hypergamy in its strict sense.

On the other hand, the fact that very high-born women are unable to ascend any further without difficulty leads to feminine encumberment of the higher social strata. This may easily be read in the percentage of women who enter the convent; the cloister thus registers the social impossibility of marriage, or the exaltation of celibacy. Marriage becomes transformed into the often sincere religious vocation of those who cannot enter into it fully in its everyday prosaic and secular form: out of 369 daughters of dukes and peers in the seventeenth century, 105 became nuns.

Women tend to move up the social scale, or to conserve their status, or to move tangentially towards convent life; on the other hand, they show an enormous repugnance towards social descent; out of 100 daughters of dukes and peers married in the seventeenth century, only five marry (horrors!) into the Robe (Labatut 1972: 138ff., 185).[9]

On a strictly qualitative level, those enjoying higher privileges are perfectly well aware of female hypergamy. Saint-Simon's wife and mother come (at different distances) from the Robe of Finance. Madame's and the Little Duke's imprecations against la Maintenon ("fairy," "old slops" or "old hag") and against the princely or quasiroyal marriages scandalously arranged by Louis XIV's bastards bear witness, in their own way, to the "resistance" faced by upwardly mobile and richly dowered women (at least, in the case of illegitimate emerging through marriage from bastard – hence inferior – status towards a quasi-supreme rank). To conclude, hypergamy in its various forms is considered inevitable; even those who constantly spit upon it must, in the end, resign themselves to it: Saint-Simon is to marry the granddaughter of the financier Frémont; the Lorge family, of excellent aristocracy and very close allies of the Little Duke, will espouse daughters of the name of Mesmes and Chamillart who, by their families, come from the Robe and the mandarin realms of bureaucracy or the ministries. Higher ranks accept wives from lower social origins; such acceptance, on the other hand, remains non-reciprocal (here I paraphrase Louis Dumont). Rank is absolutely assured by masculine transmission: this effectively one-way system of hypergamy is thus fully legitimate, and does not offend the purity of the hierarchy.

How does one escape from the oppressive system of rank or, elsewhere, from the caste system? Louis Dumont proposes the personage of the renunciator and the group constituted by the sect. Both act, outside the world, to restore or affirm individualism by comparison with the weight of corporate society which, in the temporal (and most "worldly" sense of the term), exalts the values of hierarchy, and of incorporation, of separation between ranks.

Madame throws little light on this subject, and yet her wishes to rediscover Nature set us on the right track, together with the Germanic and nostalgic souvenir of "cherries eaten on the mountain at five o'clock in the morning" and, lastly, her perpetual references to the Abbey of Maubuisson. Madame is forever wondering whether she should retire to this monastery (where one of her relations is abbess); such a retreat would help her to flee from the inconveniences of the court. On other occasions, the Palatine's question is posed in the opposite sense: should she not resign herself to the confusion, complications, and traps of Versailles or Marly? To be or not to be? This question will be resolved upon the death of Monsieur; Madame, at last widowed, will soon burn the correspondence between her husband and his favorites, and she will cry out everywhere, in stentorian tones: "The convent . . . never! Let no one speak to me of the convent."

Saint-Simon's analysis deals with this question in greater depth. From the first volume of his Memoirs (1692), he describes the solitary inhabitants of Marlagne, in the south of Namur, as being in a world apart. These recluses live in a cloistered dereliction of almost complete isolation, shut off by huge walls. They quench their thirst with pure water, and their little gardens provide sustenance. They apply themselves to physical work, silence and prayer.

> Marlagne is a monastery on a small and pleasant outcrop in a beautiful forest. It is entirely surrounded by tall trees, with a large park, founded by the archduke and duchess, Albert and Isabelle, for a Carmelite retreat of the sort which these monks have in each of their provinces, where those of their order may retreat from time to time, for a year or two, but never more than three, with the permission of their superiors. There they dwell in perpetual silence, in the poorest of cells, which are more or less akin to those of the Carthusians, though they take common meals in the refectory, which are very frugal, and exist in a state of almost perpetual fasting, strictly observing office, and sharing their time between manual work and contemplation. They have four little rooms, a little garden and a little chapel each, with the greatest abundance of the most beautiful and best springs from which I have ever drunk, both in their dwelling-places and in and around their park, most of these springs gushing freely forth. The park extends over high and low ground, with many very tall trees, and is enclosed by walls. It is extremely large. Within it there are eight or ten cottages, each far from the others, divided like those of the cloister, with a slightly bigger garden and a little kitchen. In each of these lives a monk from the house, although he rarely stays for more than one month, having retreated with the permission of his superior, who is his only occasional visitor; life in these cottages is more austere than in the house, and led in total seclusion. All attend the service on Sundays, and take away their provisions from the convent, preparing their food alone during the week. They never leave their little dwellings; upon saying mass, they ring a bell which is heard by the neighbor, who comes to give the responses, and returns without a word. Prayer, contemplation, work and simple household tasks, and basketmaking occupy their time, as in the Greek monasteries in former times.

In this same passage from the beginning of the Memoirs dealing with the military campaign of 1692, Saint-Simon contrasts the recluses with the Jesuits and their trickery, the latter preferring politics to renunciation:

> After the taking of Namur, an event occurred there which created quite a scandal, and its results might have proved vexatious for any prince other than the King. Since it would not have been fitting for the King to be in Namur during the siege of the castle, we visited all with exactitude – albeit by capitulation – prior to his entry to the town; the mines, the magazines, everything, in short, was shown to us. When we made a last visit after capturing the castle, and wished to continue our search at the Jesuits' abode, they opened up to us although they looked surprised, to say the least, at our not having trusted their testimony. However, when we had carried out a search in all the most unexpected places, we found that there were stocks of powder underground, which the Jesuits had carefully refrained from mentioning: what they were intending to do with it remains uncertain. The powder was taken away and, since those in question were Jesuits, nothing came of the incident.

The contrast is striking and it is to be found throughout the Memoirs. On the one hand there are those who retreat from worldly existence and from the society of rank – Saint-Simon periodically goes on retreat with these persons or groups, seeking friendly counsel through their contact, visiting one or another of the recluses, before whom he gapes with admiration. This applies to the monastery of La Trappe, and to Rancé, prior of this convent, or to the Comte de Charmel, who has given up court life and makes long stays at La Trappe (Boislisle V, 380ff.). Similarly there are various resident bishops and ecclesiastics, who abandon the monarchic entourage for their religious circumscriptions, and conserve a baptismal innocence looking after the poor, fasting, and aspiring to vegetarianism; at every field they cry; "Ah, my vegetables, my beloved vegetables!" On the other hand, Saint-Simon is enormously mistrustful of pseudomystics or those pious persons who are not really detached from the world, and whose aims are more political than religious; these are the Jansenists (for whom his sympathy is far from complete), the Quietists and other nuns and monks of Fénelon's type (Fénelon himself being hungry for power), or, lastly, the Jesuits, up to their necks in worldly and world-wide politics.

The model I have tried to present in the course of these pages is, as you may see, very close to that proposed by Louis Dumont with regard to India in his *Homo Hierarchicus* (Paris, 1966). Of course I am not pretending that the subculture of the court of Louis XIV derives from the civilizations of classical India, nor vice-versa. I am simply posing, in these two cases, the logic of an aristocratic system (pushed to its extreme amongst the Indian castes; already well affirmed in the society of rank whose culminating point, itself stratified, is situated at Versailles). In both cases, this logic produces homologous configurations: the exaltation of graded hierarchy extends to the individual himself; the separation between the various levels is more or less strict; the values or counter-values flee to the extremes; the poles of a double scale arise in contrast, sacred opposed to profane, impure against pure. The divisions, factional if not political, correspond to the hierarchical axes. The hypergamy of the women tempers the rigidity of rank, which descends in the male line. The image of the Renouncer, ascetic and individualist, stands out against the prevalence of worldly hierarchy.

Appendix. Dramatis personae in order of appearance

Louis de Rouvroy, Duc de Saint-Simon (1675–1755), author of the famous Memoirs, the "Little Duke."

Elisabeth-Charlotte of Bavaria, Duchesse d'Orléans, called the "Palatine Princess," "Madame," second wife of "Monsieur" (Philippe I, Duc d'Orléans, brother of Louis XIV), daughter of the Palatine Elector, Charles-Louis of Bavaria.

her son: Philippe II, Duc de Chartres, later d'Orléans (1674–1723); "grandson of France," Regent after the death of Louis XIV.

his son: Louis, Duc de Chartres (1703–52), first prince of the blood.

Duc du Maine, Louis-Auguste de Bourbon (1670–1736), son of Louis XIV and Madame de Montespan, married Anne-Louise-Bénédicte de Bourbon-Condé, daughter of Henri-Jules de Bourbon, Prince de Condé, first prince of the blood.

Françoise-Marie de Bourbon, called "Mademoiselle de Blois II" (1677–1749), daughter of Louis XIV and Madame de Montespan, wife of Phillipe II d'Orléans.

Dauphin (1713), Charles, Duc de Berry (1686–1711), brother of Louis, Duc de Bourgogne (1682–1712), third son of Louis, the Grand Dauphin (1661–1711), grandson of Louis XIV.

Louis-Françoise de Bourbon, "Mademoiselle de Nantes," daughter of Louis XIV and Madame de Montespan, wife of Louis de Bourbon-Condé, prince of the blood.

Léopold, Duc de Lorraine, husband of Mademoiselle de Chartres, daughter of the Palatine Princess.

Marquise de Maintenon, secret spouse of Louis XIV.

Françoise de Rochechouart-Mortemart, Marquise de Montespan, Louis XIV's favorite (1667–78).

Mademoiselle de la Vallière, Louise-Françoise de la Baume-Le Blanc, created Duchess by the King, royal favorite 1661–7. After her separation from Louis XIV she became a Carmelite nun under the name of Louise de la Miséricorde.

Figure 3.2 The King's legitimate family

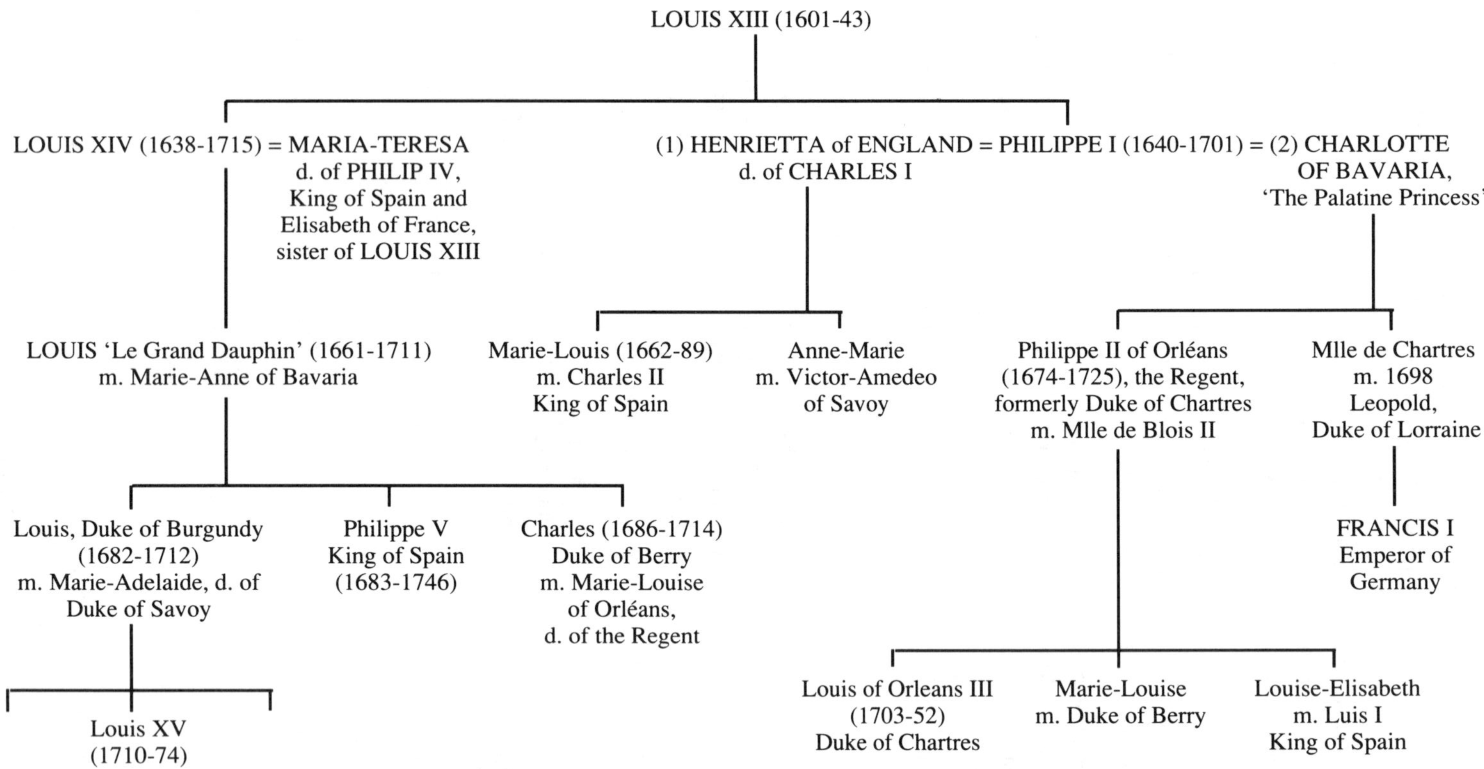

Genealogical chart adapted from LABATUT, JEAN-PIERRE: *Louis XIV, Roi de Gloire*, Imprimerie Numale, 1984, pp. 136-7

Figure 3.3 The King's "second family"

Louis XIV = (1) Mlle de LA VALLIÈRE, created Duchess of La Vallière and of Vaujours

MARIE-ANNE DE BOURBON (1666-1739)
m. LOUIS ARMAND DE BOURBON,
PRINCE DE CONTI

COUNT DE VERMANDOIS (1667-83)
Admiral of France
(died *aet.* 16)

= (2) FRANÇOISE DE ROCHECHOUART-MORTEMART, wife of Marquis de MONTESPAN, by whom she had a son, DUKE D'ANTIN and a daughter

LOUIS-AUGUSTE DE BOURBON,
DUKE DU MAINE (1670-1736)
m. ANNE-LOUISE DE BOURBON-CONDÉ,
d. of 'Monsieur le Prince' and gd. d. of the great
CONDÉ

LOUISE-FRANÇOISE,
'Mlle DE NANTES'
m. 'Monsieur le Duc'
gd. son of the great
CONDÉ

FRANÇOISE-MARIE,
'Mlle de BLOIS II'
(1677-1749) also
called 'Madame Lucifer,'
m. Duke of CHARTRES,
later the Regent

MARIE-LOUISE (1695-1719)
m. Duke of BERRY

COUNT OF TOULOUSE
(1678-1737)

= (secretly married) FRANÇOISE D'AUBIGNÉ, widow of PAUL DE SCARRON, poet and playwright, governess to the children of Madame de Montespan before succeeding her. No issue

Her daughter, Marie-Anne de Bourbon, Mademoiselle de Blois I, legitimized by royal decree, like the king's other bastard children, around the ages of three or four. (It was hardly worth doing this earlier considering the considerable risk of infantile mortality, to which succumbed four of her mother's eight children by Louis XIV.)

Louis-Joseph de Bourbon, Duc de Vendôme, legitimized great-grandson of a bastard of Henri IV, married to Marie-Anne de Bourbon-Condé, called "Mademoiselle d'Enghien," whose rank as princess of the blood was confirmed by Louis XIV.

Duc d'Antin, Louis-Antoine Pardaillan de Gondrin, legitimate son of Marquis and Marquise de Montespan, created Duc d'Antin by Louis XIV.

Gaston d'Orléans, brother of Louis XIII.

Granddaughters of France: Anne-Marie d'Orléans, Duchesse de Montpensier, called "la Grande Mademoiselle," and Mademoiselle d'Alençon, Duchesse de Guise.

Duchesse de Choiseul, Louise-Gabrielle de La Vallière, cousin of the Princesse de Conti.

Louis XV, second son of Louis, Duc de Bourgogne, great grandson of Louis XIV.

Louis-Alexandre de Bourbon, Duc de Toulouse, second son of Louis XIV and Madame de Montespan.

Notes

1 References are to the Boislisle edition of Saint-Simon's Memoirs. Details of aristocrats mentioned in this chapter are given in the Appendix, pp. 73–6.
2 Translator's note: The French term "carrosse cloué" refers to the privilege of having the protective covering on one's coach nailed in place, rather than simply suspended from hooks (as, for example, is the case for Duchesses).
3 Illegitimate son of Louis XIV, as was his brother, the Comte de Toulouse. His wife, the Duchesse du Maine, is of the Condé family and thus, as princess of the blood, is of higher rank than her husband whose rank is theoretically "diminished" by his bastardy.
4 Translator's note: The "affaire du bonnet" was one of a series of attempts by the Parliamentary presidents, and especially by Novion, to avoid according any distinction to the dukes and peers. Normally, upon calling for the "ayes" and "nays" of members of Parliament, the President was to doff his mortar on addressing the peers and then the princes of the blood. Novion, however, became increasingly negligent in the removal of his mortar, to the point where the peers found themselves regularly addressed by a "covered" president.
5 Translator's note: A small table at the side of the communion table on which the bread and wine are placed before consecration.

6 Cf. also Boislisle II, 259: the Dukes were excluded not from Holy Communion, but from adoration of the Cross.

7 See also the offertory-bag affair; this was a strike of the collection normally taken up by the princesses so as to further lower the duchesses, obliged to take up the collection (Boislisle XI, 354).

8 Impurity for our authors concerns, above all, sex and the transmission of life (bastardy, sodomy, syphilis), while in India it extends to all kinds of domains: excrement, the smoking of another person's pipe, etc. Without prejudicing the Indian case, we may say that our French authors believe very strongly in the heredity of acquired characteristics, in the broad and at times extravagant sense of this expression. The following text written by Madame (August 14, 1718), and which has its equivalents in Saint-Simon, illustrates this point:

> At that time I rendered your mother an important service. When she was pregnant with Charles-Maurice, His Grace my father, wanting to give her a letter for me while she was in bed, moved too brusquely and struck her, thus injuring the eye which became swollen, and the next day was black and blue. Seeing her thus disfigured, I was afraid and said to her: "Lord Jesus, Madame, what an eye!" For her peace of mind, she told me how this had come about. When Charles-Maurice was born, he had one eye which was as though completely out of the socket. You know, dear Louise, our father the Elector was horribly jealous; he imagined that Madame our mother had too often looked at Colonel Webenheim, who had but one eye and who came to play with us on so many occasions, and he thought that this was the reason that Charles-Maurice had an eye as black as the Colonel's patch. Disturbed, he called me upon the birth of the child and said "Liselotte, you see that eye, is it not black as Colonel Webenheim's patch?" I began to laugh and said "No, Your Grace, I know where that comes from!" The Elector, in a temper, cried "By the Sacrament! Whence then does it come?" "It is something Your Grace did not see; you remember when, during the voyage to Openheim, you wished to give madame a letter from my mother telling her to keep me the next day, and you struck her on the eye? The following day, it was as black as the child's eye now is." "My God," said the Elector, "How relieved I am that you remembered such a thing! For the love of God, say nothing of this to Madame."

In these conditions, one can understand that a traumatism or deviance of sexual life may be considered a danger to the purity of the lineage (itself assigned to a particular rank). If one is to believe even the superior minds of the seventeenth century, the process of transmission of life is seen to be terribly fragile, and subject to all possible chance events.

9 I.e., the "noblesse de robe": descendants of those lawyers and administrators ennobled by a grateful sovereign, but of more recent origin than the "noblesse d'épée," the nobility of feudal origin. [Eds.]

References

Boislisle, A. de (ed.) 1879–1930. *Duc de Saint-Simon, Mémoires*, 43 vols., Paris
Debray Régis 1981. *Critique de la raison politique*, Paris
Dumont, Louis 1967. *Homo Hierarchicus*, Paris
Elias, N. 1974. *La Société de cour et la dynamique de l'Occident*, 2 vols. Paris
Hazard, P. 1935. *La Crise de conscience europienne*, 2 vols., Paris

Labatut, J.-P. 1972. *Les Ducs et pairs de France au XVII siècle*, Paris
Leroy Ladurie, Emmanuel 1972. *La Territoire de l'historien*, 2 vols., Paris
Saint-Simon, see Boislisle
Van der Cruysse, D. 1971. *Le Portrait dans les mémoires de Saint-Simon*, Paris

4

Rites as acts of institution

PIERRE BOURDIEU
(translated by Roger Just)

In an article entitled "Educational tests and social consecration"[1] I have shown not only how formal education produces technical results (inculcating both learning and learning how to behave) but how it possesses a truly magical effect of initiation and consecration which is clearly seen in the case of "schools for the elite": British public schools, the boarding schools reserved for Samurai, the sacred colleges of the Maori or the preparatory courses for the French Grandes Ecoles, that is to say, the institutions devoted to training those who are destined to become members of the dominant class. The Grandes Ecoles – Ecoles Normales Supérieures, Ecole Nationale d'Administration, Ecole Polytechnique, Ecole Centrale, Institut National Agronomique, Ecole des Hautes Etudes Commerciales, etc. – are institutions of higher education which have only a small number of pupils in comparison with the universities. The students gain admission through a *concours* (which is a competition rather than an examination) for which they are prepared during two or three years after the baccalaureat in those coaching courses for the Grandes Ecoles which are known as "Khâgne" and "Taupe": Khâgne for the humanities, Taupe for the sciences. From those who have passed their "bachot" the best are offered the possibility of continuing their studies in the Lycée under a separate body of teachers (the most prestigious) in anticipation of the *concours* which will determine their career for the rest of their life.

As was shown by a statistical study of the scholarly and social qualities of their students – and all the more so in the case of the students of the Grandes Ecoles themselves, who have been through a second process of training and selection – the training provided by these institutions succeeds so well only because it is applied to students already prepared in accordance with the demands of the training operation. The Grandes Ecoles in fact skim off from the schools the elite, which is also a social elite: 60–70 per cent of their students come from the dominant class as against 30–40 per cent of university students. It all happens as though the educational system, in the manner of Maxwell's demon,[2] channeled towards each of the Grandes Ecoles and hence towards the fraction of the dominant class to which it provides the entry the individuals best equipped with the dispositions which it is supposed to inculcate and who are the product of the characteristic education of these fractions. Hence the sons of professors and school masters are directed mostly to the Ecoles Normales Supérieures which prepare the teachers of higher education and researchers; the sons of the higher civil servants are sent to the Ecole Nationale d'Administration, the source of the higher posts in government organizations; and to the Ecole des Hautes Etudes Commerciales which fills the posts of command and higher administration go the sons of the chiefs of industry and commerce and so forth.

The study of the Grandes Ecoles and their preparatory courses also showed, once analyzed, the technical effects produced by the educational organization of these establishments, in particular the encouragements, constraints, and continual controls which are exercised in these institutions with the aim of reducing the lives of the students to an uninterrupted succession of scholastic activities and which are aimed, in the cause of what one might call a "culture of precipitation," to instill the capacity to "collect one's ideas rapidly" and to "deal creditably with any problem that arises." The entrance examinations are a way of imposing a kind of *numerus clausus*, an act of separation and closure which installs between the last to be accepted and the first to be refused a social chasm. And the process of transformation which the Grandes Ecoles accomplish is quite analogous to that which is instituted by rites of passage: by means of the magical operations of *separation* and *aggregation* analyzed by Van Gennep it tends to produce a consecrated elite, that is to say not only distinct and separate, but also recognized and recognizing itself as worthy to be so – in a word "distinguished."[3]

With his notion of a rite of passage, Arnold van Gennep named, indeed described, a social phenomenon of great importance. I do not believe that he did much more. Nor has any great advance been made by those who, like Victor Turner, have resuscitated van Gennep's theory and put forward a more explicit and systematic description of the *stages* of ritual. In fact it seems to me that, to go any further, certain questions have to be asked which the theory of the rite of passage does not ask, and in particular, questions which concern the social function of ritual and the social significance of the line, the boundary, whose crossing, whose transgression, the ritual renders legitimate. For it is questionable whether, by putting the emphasis on temporal passage (from childhood to adulthood, for example), the theory might not be masking one of the essential effects of the rite: to separate clearly those who have undergone it not from those who have not yet undergone it, but from those who will under no circumstances undergo it, and thus to institute a lasting difference between those whom the rite concerns and those whom it does not concern. It is for this reason that I would prefer to speak not of rites of passage, but of rites of consecration, or of legitimization, or quite simply of *rites of institution* (giving to this word the active sense which it has, for example, in the expression "the institution of an heir"). Why should we substitute one word for another? Let me invoke Poincaré, who defined mathematical generalization as "the art of giving the same name to different things". Poincaré insisted on the critical importance of the choice of words. When language has been well chosen, he said, demonstrations made about a known object apply to all sorts of new objects. The analyses which I shall advance have been produced by generalizing the results of my analysis of the functioning of the elite schools. I should like to engage in the somewhat perilous exercise of trying to isolate the invariant properties of social rituals understood as rites of institution.

To speak of rites of institution is to point to the fact that every rite leads towards the consecration or legitimization of an arbitrary boundary, that is to say, it attempts to misrepresent the arbitrariness and present the boundary as legitimate and natural; or (and this amounts to the same thing) it effects a solemn, that is to say a licensed and extraordinary, transgression of the boundaries which constitute the social and ideational order which it is concerned at all costs to protect (as in the separation of the sexes in the case of marriage rituals). By solemnly marking the passage across a line establishing a fundamental division in the social order, the rite attracts the observer's attention towards the passage (hence the expression "rite of passage"), whereas what is important is the line. And what does this line separate? A before and an afterwards, to be sure: the non-circumcised child and the circumcised child, or even the set of uncircumcised children and the set of circumcised adults. In fact the most important division, and the one which passes unnoticed, is that produced between the set of those who can be subjected to circumcision (boys and men, regardless of whether they are children or adults), and those who cannot (girls and women). There is therefore a hidden set by reference to which the constituted group is defined. The major effect of the rite is the one which goes completely unnoticed: by treating men and women differently the rite consecrates the difference, it institutes it – and at the same time it institutes man *qua* man, that is to say, circumcised, and woman *qua* woman, that is to say incapable of undergoing this ritual operation. An analysis of Kabyle ritual reveals this clearly: circumcision separates the young boy not so much from his childhood or from boys who are still children, as from women and the feminine world, that is to say, from his mother and from all that is associated with her: the wet, the green, the raw, springtime, milk, blandness, etc. We might note in passing that just as the process of institution consists in assigning properties which belong to social nature but which are destined to appear as properties of natural nature, so, as Pierre Centlivres and Luc de Heusch have observed, rites of institution also lead logically towards an integration of oppositions which are properly social (such as masculine and feminine) with a series of cosmological oppositions (such as man is to woman as sun is to moon). This form of integration provides a very effective means of naturalizing social properties. Thus rites which are sexually differentiated consecrate the difference between the sexes. They constitute as a lawful distinction, as an institution, what is a simple difference of fact. The separation brought about by the ritual (which itself enacts a separation) brings about a consecration.

But do we really understand what it means to consecrate something and to consecrate a difference? How is what I shall call the "magical" consecration of a difference brought about, and what are its practical consequences? Is it possible that the fact of socially instituting a preexistent difference (such as that which distinguishes the sexes) by an act of constitution has only a symbolic effect (in the sense given the term when we speak of a symbolic gift), that is to say, none at all? The Romans had a saying: "You're teaching a fish how to swim." That is indeed what the ritual of institution does. It says: "This man is a man" – meaning by that "he is a real man," which is something a little more. It strives to make of the smallest, the weakest, in short the most effeminate man, a man who is fully man and separated by a difference of nature, of essence, from any woman however masculine, large, strong, etc. In this case to institute is to consecrate, that is to say to sanction and to sanctify, a state of affairs, an established order – which is precisely what, in the juridico-political sense of the word, a constitution does. Investiture (of a member of the Légion d'honneur, of a Deputy, of the President of the Republic) consists in sanctioning and sanctifying, in making known and recognized, a difference (preexistent or not); in making it exist as a social difference known and recognized by the agent thereby invested, and by others. In short, at the risk of hindering an understanding of the most basic social phenomena, and in pre-capitalist societies as well as in our own (a diploma is as much a piece of magic as are amulets), the social sciences must take account of the symbolic efficacy of rites of institution, that is to say, of the power they have to act on the real by acting on the representation of the real. Investiture, for example, employs a symbolic efficacy which is completely real in that it does, in reality, transform the person who is consecrated – first, because it transforms the representations which other agents make of him, and perhaps above all the attitudes which they adopt towards him (the most evident of these changes being the fact that he is granted titles of respect, and the respect which is actually associated with their enunciation); further, because it transforms at the same time the representation which the person who has been invested makes of himself, and the attitudes which he believes it is necessary for him to adopt in order to conform to this representation. We could include within this same logic the effect of all titles of credit and credence – the English call them credentials – which, like aristocratic or academic titles, lastingly increase the value of their bearers by increasing the breadth and intensity of belief in their value.

The process of institution is an act of social magic which has the capability of creating differences *ex nihilo*, or equally (and this is more frequently the case) of exploiting in some way or other preexistent differences, such as the biological differences between the sexes, or, for example, in the case of someone who inherits by primogeniture, differences of age. In this sense it is, like Durkheim's notion of religion, "a well-founded aberration," an act of violence which is symbolic, but *cum fundamento in re*. The most socially effective distinctions are those which give the appearance of being based on objective differences (I am thinking, for example, of the notion of a "natural frontier"). It remains true, as can be clearly seen in the case of social classes, that one is almost always dealing with continuums, with continuous distributions, since different principles of differentiation produce different divisions which are not always completely superimposable. Social magic, however, always manages to produce discontinuities out of the continuity. The example *par excellence* is the competitive examination which first prompted me to think about the problem. Between the last person to get through it and the first person to be scratched, the exam creates an all-or-nothing difference which will last for the rest of the candidates' lives. One will become a "polytechnician" with all the attendant advantages; the other becomes a nonentity. None of the criteria which can be adduced to justify technically the nobility's ditinction (as a legitimate difference) fits this case. For example, the most wretched swordsman from the nobility remains noble (even if, according to different national traditions and epochs, his image is thereby somewhat tarnished). Conversely, the best swordsman from amongst the commoners is still a commoner (even if he manages to derive a certain "nobility" from his excellence in a typically "noble" field). And the same can be said for each of the criteria which, at any given time, define the nobility; bearing, elegance, etc. The institution of an identity, whether a title of nobility or a stigma ("you're nothing but a . . . "), is the imposition of a social essence. The process of institution, the assigning of an essence, a competence, is the imposition of a right to be, which is at the same time the imposition of an expectation and even a duty to be, something. It is to signify to someone what he is, and to signify to him that he must comport himself in a certain manner in consequence. The indicative is in this case an imperative. The ethic of honor is only a developed form of the formula which consists in saying of a man that "he is a man." To institute, to give a social definition, an identity, is also to impose certain limits, and *noblesse oblige* could be rendered as Plato's *ta heautou prattein*, to do what it is one's essence to do, and not something else – in a word, for a nobleman it is not to degrade himself, to uphold his place. It befits noblemen to act nobly, and the principle of nobility can be seen in the noble action just as much as the principle of noble action in the nobility. This morning I read in the newspaper: "It falls to the President of the Federation, Kurt Furgler, to express on Tuesday evening the condolences of the Federal Council to the Egyptian people following the death of Anouar Sadat." The authorized spokesman is he to whom it falls, whom it behoves, to speak in the name of the collectivity. It is simultaneously his privilege and his duty, his proper role, in short, his competence (in the legal sense of the term) to do so. Social essence is the set of these attributes and of these social attributions which the act of institution produces as a solemn categorization which tends to produce what it designates.

The act of institution is therefore an act of communication, but of a particular kind: it signifies to someone his identity, but at the same time as it expresses that identity and imposes it on him, it expresses it before everyone (*kategoresthai* originally meant to accuse publicly) and authoritatively informs him of what he is and what he must be. This is readily seen in insults, which are a sort of malediction (*sacer* also means cursed), which tend to enclose their victim in an accusation which acts as a destiny. But it is even truer of investiture or of nomination, judgments of true social attribution, which assign to their objects everything inscribed in a social definition. It is through the results of ascription (*noblesse oblige*) that the ritual of institution produces its most "real" effects. He who is instituted feels himself summoned to conform with his definition, with the loftiness of his office. The heir who has been designated (by a more or less arbitrary criterion) is recognized and treated as heir by all his group, primarily his family, and this different and distinctive treatment cannot but encourage him to realize his essence, to live in conformity with his social nature. The sociologists of science have established that the greatest scientific advances have been made by researchers who have come from the most prestigious scholarly institutions. This can be explained in large part by the higher level of subjective aspiration which determines the collective, that is to say objective, recognition of their aspirations and their assignment to a class of agents (men, students of the Grandes Ecoles, established writers, etc.) to whom such aspirations are not only assigned and credited as rights and privileges (in contrast to the pretentious claims of other aspirants) but also assigned and imposed as duties by constant backing, encouragement, and calls to order. I am thinking of that drawing by Schultz in which Snoopy is shown perched on the roof of his kennel saying, "How can you be modest when you know you're the best?" It ought simply to read: when it is a matter of public fact – the result of official recognition – that one is the best, *aristos*.

"Become what you are." Such is the formula which underlies the performative magic of all acts of institution. The essence assigned by nomination, by investiture, is, in a true sense, a *fatum*. (This applies also, and especially, to those injunctions, sometimes tacit sometimes explicit, with which members of a family group continually admonish children, and which vary in intention and intensity according to social class and, within any class, according to sex and rank within the group.) All social destinies, whether they are positive or negative, blessings or stigmas, are equally *fatal* (by which I mean mortal) because they enclose those whom they distinguish within assigned boundaries which they are made to recognize. The self-respecting heir behaves like an heir, and will become heir to his inheritance in accordance with Marx' formula – that is to say, he will become himself invested in things and appropriated by the things which he has appropriated. There is always the possibility of some accident, of course: the unworthy heir, the priest who defrocks himself, the nobleman who degrades himself, the bourgeois who goes to the dogs – these exist. But a limit, the sacred frontier, is still encountered. Owen Lattimore said of the Great Wall of China that it was built not only to keep foreigners out, but also to keep the Chinese in. This too is the function of all magical frontiers (whether we are talking about the frontier between masculine and feminine or between the elect and the excluded of the academic system): to prevent those who are inside and on the right side of the line from going over, from losing caste, from lowering themselves. Elites, said Pareto, are destined to extinction once they cease to believe in themselves, once they lose their morale and their morality and are prepared to cross over the line in the wrong direction. This too is one of the functions of acts of institution: to discourage permanently the temptation to cross over, to transgress the boundary, to desert, to resign.

Every aristocracy has to expend a considerable amount of energy in persuading its elect to accept the sacrifices which are involved in privilege or which are involved in the acquisition of those lasting predispositions which are the condition of retaining privilege. When culture is representative of a dominant class (that is to say, almost always) which tends towards asceticism, tension, and contention, the task of institution must be to reckon with the temptations of nature or of counter-culture. (I should point out in parenthesis that in speaking of the task of institution, and in making the more or less painful inculcation of lasting predispositions an essential component of the social operations of the process of institution, I am doing no more than giving the word "institution" its fullest meaning. Having recalled, with Poincaré, the importance of choosing one's words, I think it not a waste of time to point out that the different meanings of *instituere* and *institutio* have only to be assembled to obtain the idea of an inaugural act of constitution, of foundation, indeed of invention, which leads via education to the possession of lasting predispositions, habits and customs.) The strategy universally adopted to combat any temptation not to behave according to one's rank consists in naturalizing difference and in turning it into a second nature through its inculcation and incorporation as a set of habits. It is in these terms that one can explain the role given to ascetic practices, and even to the physical suffering which occurs in all negative rites destined, as Durkheim says, to produce men who are out of the common, in a word, distinguished) and also in all the apprenticeships which are universally imposed on the future members of an elite (an apprenticeship in dead languages, a prolonged period of seclusion, etc.). All groups entrust to the body, considered as a memory, their most precious deposits. And the use which all rites of initiation in every society make of physical suffering becomes comprehensible when it is known (as psychological experiments have proved) that the more severe and painful the rites of initiation, the more people adhere to an institution. The purpose of inculcation, which results in the lasting imposition of arbitrary boundaries, is to naturalize those decisive breaks which are constitutive of cultural arbitrariness (those expressed by the fundamental oppositions of masculine/feminine, etc.) by translating them into self-conscious limits, which incline some people to uphold their station and keep their distance, and others to keep their place and be content with what they are, with what it is their place to be – thus depriving them even of deprivation. It can also lead to the inculcation of lasting predispositions such as those of class taste, which, since they are based on the principle of a "choice" of exterior signs whereby social position is expressed (like clothes, but also mannerisms or language), means that all social agents are the bearers of distinctive signs (of which signs of distinction are only a sub-class) designed to unite them or separate them as surely as physical barriers or explicit prohibitions – I think, for example, of class homogamy. Those signs which are embodied in everything we call manners (manners of speech, i.e., accents; manners of walking or of holding oneself, i.e. gait, bearing, comportment; table manners) or in taste (the principle of production of all practices which are destined, wittingly or unwittingly, to signify social position by playing on distinctive differences) are destined to function, and more effectively than external bodily signs (like decorations, uniforms, braid, insignia, etc.), as so many calls to order by which those who have forgotten it, who have forgotten themselves, are reminded of the place which institution assigns them.

The power of the categorical judgments of attribution which the process of institution effects is so great that it is capable of withstanding all practical contradictions. Kantorowicz's analysis of the "King's two bodies" is well known: the person of the invested king survives the person of the biological king which is mortal and subject to imbecility, or sickness, or death. In the same way, if a polytechnician turns out to be hopeless at mathematics, people will assume that he has deliberately chosen to appear so, or else that he has directed his intellectual energies towards more important matters. But the best illustration of the autonomy of ascription in relation to achievement (for once Talcott Parsons can be invoked), of social being in relation to doing, is without doubt furnished by the fact that strategies of condescension can be resorted to whereby one's social definition may be substantially contradicted whilst nevertheless remaining the basis of the way one is perceived. I term strategies of condescension those symbolic boundary transgressions which allow one simultaneously to profit from conformity with a definition and to profit from its transgression. Such is the case when the aristocrat slaps his groom on the bottom. People say, 'He's a simple man" – it nevertheless being understood that he is a simple man for an aristocrat, that is to say, for a man of superior essence whose essence, in principle, is not compatible with such an action. Actually the matter is not so simple, and a distinction ought to be introduced here. Schopenhauer speaks somewhere of the "pedantic comic," that is to say, of the character who raises a laugh by performing an action which though natural for him, is not in keeping with the idea we have of him – "like a stage horse", says Schopenhauer, "who suddenly starts to drop dung." Schopenhauer is thinking of professors, like Professor Unrat in *The Blue Angel*, of whom we have so strongly and rigidly defined an idea that any transgression of its limits is clearly evident. In contrast to Professor Unrat, who is carried away by his passion and loses all sense of the ridiculous, or of dignity (which amounts to the same thing), the condescending member of the elite deliberately chooses to go over the limits. He possesses that privilege of privileges, which consists in taking liberties with his privilege. Thus in matters of linguistic usage, the bourgeois, and above all intellectuals, can allow themselves to employ forms which are less than proper, which are lax, and which are forbidden the *petits bourgeois* who are condemned to the ultra correct. In short, one of the privileges of consecration lies in the fact that by conferring on the person it consecrates an indisputable and permanent essence, it authorizes him to make transgressions which would otherwise be forbidden. He who is sure of his cultural identity can play with the rules of the cultural game. He can play with fire and say that he likes Tchaikovsky or Gershwin or even (a piece of audacity) Aznavour or B movies.

Acts of social magic as different as marriage or circumcision, the conferment of degrees or titles, the dubbing of a knight, nomination to positions, offices, and honours, the affixture of a stamp, a signature, or initials, can all succeed only if the institution, in the active sense of an act leading to the institution of someone or something endowed with such and such a status or such and such a title, is also an act of institution in another sense, i.e., an act guaranteed by the entire group or by a recognized institution. Even when it is accomplished by an individual agent duly mandated to accomplish it and to accomplish it according to recognized forms (that is to say, according to conventions held to be suitable in terms of place, moment, instrument, etc., the combination of which constitutes the certified, that is, socially valid and therefore effective, ritual), its basis lies in the belief of an entire group (which may be physically present), that is to say, in dispositions which are socially fashioned to know and to recognize the institutional conditions of a valid ritual (which implies that the symbolic efficacity of the ritual will vary, simultaneously or successively, according to the degree to which those whom it addresses are more or less prepared and disposed to accept it). This is what is forgotten by linguists in the tradition of Austin who look to words themselves for an "illocutionary force," and which they isolate at times as "performatives." In contrast to the imposter who is not what one takes him to be, who, in other words, usurps the name, title, rights and honours of another, and in contrast also to the mere "stand-in," a deputy or assistant who plays the role of director or professor without having the titles, the legitimately mandated agent, for example an authorized spokesman, is an object of guaranteed credence, certified as genuine. He is what he appears to be. He is really what each person believes him to be, because his reality (whether of priest, professor, or minister) is founded not on his own belief or individual pretension (which is always open to slight and detraction – "Who does he take himself for?", "Who does he think he is?" etc.), but in collective belief guaranteed by institution and made concrete by a title or by braid, uniform or other symbols of office. Manifestations of respect, which consist for example in giving someone his title (Mister President, Your Excellency, etc.), are so many repetitions of the inaugural act of institution which was performed by a universally recognized authority, and which was therefore founded on the *consensus omnium*. They are the equivalents of an oath of allegiance, a sign of recognition towards the particular person whom they address, but above all towards the institution which has instituted him (which is why the respect of forms and the forms of respect which define polite behavior are so profoundly political). Universal credence, which preexists rituals, is the condition of ritual efficacity. One preaches only to the converted. And the miracle of symbolic efficacity disappears if one realizes that the magic of words does no more than release forces – predispositions – which have already been assembled.

In conclusion I should like to raise a question which I fear may seem a little metaphysical. Could rites of institution, whatever their form, exercise the power they do (I am thinking of the most obvious case, those decorations and distinctions which Napoleon called "fripperies") if they were not able at least to supply the semblance of a meaning, of a *raison d'être*, to those beings who have no reason to be, i.e. human beings, and to allow them the feeling of a purpose, or quite simply an importance, any importance, which might snatch them from insignificance? Doubtless the true miracle produced by acts of institution resides in the fact that they succeed in convincing those individuals whom they consecrate that they are justified in existing and that their existence serves some purpose. But by a sort of curse, the essentially diacritical, differential, distinctive nature of symbolic power means that the access of one distinguished class to Being has for its inevitable counterpart the decline of the complementary class into Nothingness, or minimal Being.

Notes

1 "Epreuve scolaire et consécration sociale", *Actes de la recherche en sciences sociales* 39 (September 1981), 3–70.

2 Translator's note: the nineteenth-century physicist Maxwell imagined a demon, capable of contradicting the laws of thermodynamics by separating the hot and cold particles each to opposed sides.

3 Translator's note: the word is used in the special sense given to it by the author in his massive work, *La Distinction*, Paris, 1980.

References

Centlivre, Pierre and Luc de Heusch 1986. *Les Rites de passage aujourdhui*, Actes du colloque de Neuchatel 1981, Lausanne

van Gennep, Arnold 1981 (1909). *The Rites of Passage: étude systématique des rites*, Paris

5

Religion, world views, social classes, and honor during the sixteenth and seventeenth centuries in Spain

JULIO CARO BAROJA
(translated by Victoria Hughes)

The complexity of life in European societies from the Renaissance down to the present makes it extremely difficult to characterize them in the sort of global terms often used in studies of primitive societies, which are assumed (though not always correctly so) to be very homogeneous and therefore easier to describe and characterize than urban, industrial societies at a more advanced stage of cultural and economic development. Historians, however, have written books describing "Renaissance Man" or "the Baroque Spirit" in very general terms, characterizing entire societies, such as France, Italy, or Spain, so that they appear much more homogeneous in their histories than they were in reality. At times this is due to a confusion between certain generalized concepts which operated at public, political, and religious levels, and a social reality which was far more multicolored and complex. It was also often contradictory or full of individual, very distinctive tonalities.

I have attempted to combat such simplifications and confusions in this essay, referring to Spain in concrete terms. The main results of my research in this field can be found in my book, "Complex Forms of Religious Life. Religion, Society and Character in Sixteenth- and Seventeenth-Century Spain."[1] There were two basic starting points to this work. On the one hand, it resulted from the need to study certain generalized religious principles, certain concepts of Christian, Catholic dogma. On the other hand, the need to examine the real social situation was equally important. For the analysis of the data shows that, whilst in some aspects there was coherence, in others contradictions arose, which were the subject of deliberation by the churchmen of the time, who tried to resolve them in their thoughts and writings. In other words, we are led to believe that the authority of Catholic theology was the decisive factor, whilst, at the same time, we are confronted with the fact that social life flowed along quite independent channels. The solution is to be found in a special activity whose importance is hard to imagine in our modern-day, laicized societies: moral theology.

We first of all encounter degrees of intensity and discernible differences in what was Godly thinking from the strictly religious point of view (the central theme). For in Hapsburg Spain there were at the same time great (and also less great) mystics and ascetics, believers with a more pragmatic idea of religion and men with only vague, unformulated ideas in this regard. But, at the other extreme, there was no lack of "atheists" (as they were then called), who were for the most part men who denied the immortality of the soul. There were also several groups with non-Christian ideas of God: descendants of Jews or Moors, who lived as they could, under the effects of forced baptisms which did not exempt them from being considered as an inferior social stratum: a source of pollution. The effects of this concept on public life were enormous. But alongside this Godly thinking, with all its varying degrees and nuances, we find another type of thought: preoccupation with the Devil. The role of the Devil in this essentially theologized society was fundamental, because if, as Saint Teresa thought, God could even be found amongst the cooking pots, then the same could be said of the Devil . . . who broke them. Fear of the Devil was as much cultivated in Spanish religious literature as was God's love and love of God. There were constant allegations of the existence of servants of the Devil. Some of these had a direct relationship with him, as, for example, witches, wizards, and sorcerers, who entered with him into an explicit pact. Others were more indirectly related, or only occasionally so. Quite apart from the sincere believers whose dominant vices led them into diabolical temptations, there were also many others who deliberately chose to be in permanent subjection to the Devil: from the popular anti-clericals who refused to respect the life and customs of the clergy, and made jokes about certain Church institutions, to atheists and unbelievers. The Catholic way of thought also considered that heretics together with the Judaisers and Islamists, belonged to this infernal army, this veritable host. At the beginning of the period, heretics were thought to have contact with Lutheran teachings: there was an abundance of Inquisitional trials against people who made statements smacking of Lutheranism during the first half of the sixteenth century. Later, in the second half, the most frequently occurring accusation was that of "Calvinism."

All this is quite sufficient to make clear that what might be called the "monolithic view" of Spanish Catholicism is a mistake. The most that could be said in its favor is that both Church and State wished to impose a "Unity of Faith." However, there is more to the issue than that. Within the ecclesiastical body itself, amongst the regular and secular clergy who formed a substantial part of Spanish society, there were areas of dispute or at least a lack of agreement, at times almost absolute, on fundamental subjects concerning the meaning and development of human life, considered both collectively and individually. The argument over the respective powers of Divine Providence as against Free Will in the evolution of Mankind is not only to be found between Protestants and

Catholics, but also within the very heart of Catholicism itself, where there were groups which denounced each others' doctrines, to the point of violent antagonism. On the other hand, until the beginning of the eighteenth century and even later, a decisive influence over this society of supposedly homogeneous ideas was exerted by a pre-Christian concept of life: a belief in astrology. Although at the end of the sixteenth century the Church publicly stated its condemnation of astrological concepts on account of their total limitation on Free Will in human actions (not for the fact that they rested on belief in the power of the stars over natural happenings), the practice of horoscope reading and astrological divination of the future continued and there were even openly astrological works written by churchmen. On the other hand this so-called science was also subject to refutation.

In the world of religious ideas, opinions about the path of humanity within Christianity show an over-riding belief in the moral decadence of present-day Man. In other ages Man was considered to have been better from all points of view. This belief, which can be found repeatedly in sermons, also had a pre-Christian root in the theories expounded by Hesiod on the different "ages" of Man: starting with the Golden Age, followed by the Silver and ending in the Iron Age, the present – tough and ferocious. Don Quixote could talk to humble, illiterate people about the "Golden Age" and be sure of being understood – it was not merely an academic concept. Similarly, the preachers could thunder against the corrupt habits of the times, combining commonplace observations on this subject with others of a more apocalyptical nature. What we might call "apocalyptical literature" had deep significance in such a social milieu and became very popular.

The "estates" constituting the society

We say "social milieu," but it would be more correct to refer to "social milieux" in the plural. Because, since the already long-past Middle Ages, social reality had been reflected in philosophy and literature as a living body, an organism comparable to the human body with all its different parts, each presupposing the simultaneous functioning of the rest. John of Salisbury seems to have been the original creator of this organic concept of society, in which each activity was represented by way of simile with the head, hands, stomach, feet, etc. From kings and princes to peasants, with soldiers, lawyers and merchants between, each was likened to an organ, with its correct place and its own corresponding function. Every "estate" had to account for its existence at the hour of death and the Last Judgment. Thus, writers of treatises on moral theology concerned with confession divided them up from the highest (popes, emperors, kings, princes) to the lowest (merchants, etc.) with a stratum of intermediaries (aristocrats, soldiers) or professional specialists (doctors, lawyers, monks) between them. Basing their work on both their own experience and other people's, these writers managed to determine which "sins" were most frequent in, or peculiar to, each "estate." It was obvious to them that, for example, there were some "sins" which were almost exclusive to great prelates and aristocrats and others, just as specifically, were peculiar to soldiers, merchants, lawyers, etc.

An extension of the same line of logic was apparent in the considerable number of oratory pieces in books of sermons, in which the division by "estate" was followed, such as those in what was called "Sacred Rhetoric" which criticized "estates" for their specific defects. There were also several oratory pieces which, on the death of a man or woman belonging to one of the "estates," who could be considered as a paragon of Christian conduct, provided an analogy of the deceased, a posthumous eulogy, proposing him as a model for others. In much the same way, other literary genera, less directly related to the Church, reflected how clear the division by "estate" must have seemed to the popular imagination, not only in Spain, but in all Western Christian countries. It was used both in literature and in the visual plastic arts. For example, in the famous "Dances of Death" and also, to some degree, in the "Ship of Fools" into which everyone entered, from the Pope to the very last usurer or "mechanical man," as those who held lowly offices or carried out tasks considered vile or infamous were called.

When we consider, on the one hand, the absolute ideal of Christian brotherhood and charity and, on the other, social reality, full of differences, categories and even, as has been seen, contradictions with regard to the nature of the world, we come up against a problem which needs clarifying. Christian morality was in conflict, not at an individual level, but at a social, collective level, each "estate" having its distinct lifestyle. This can be seen in the conflicts in the image of the peasant, the man closest to the earth; or, finally, in that of the soldier or the noble. Each one had its own idea about honor.

The double-sided image of the peasant

The image of the peasant given in sixteenth- and seventeenth-century texts is contradictory. There were not usually any chapters in confessional manuals devoted to analyzing which sins were peculiar to this "estate." However, in many other texts such an analysis paints him in contrasting lights. The villein, i.e. the humblest of peasants, was thought to be prey to all the worst passions. From this arose the concept of "villainy"[2] which made the connection between "villeinage," defined as "lowliness of birth, condition, or estate," and the idea of "villainy" as "a vile, despicable action" considered peculiar to rustics, according to the *Diccionario de Autoridades* of 1739. In general, there were a high number of proverbs and sayings in which the villein appeared as the symbol of all that was lowly and abject, as the man lacking in any idea of honor, as a traitor, opportunist, etc. Villainous actions were considered to belong to the villein in much the same way as his food, clothes, and language.

"Give a villein your foot and he'll take your hand"; "Put a villein on a mule and he won't know either God or the world." However, the word "rustic" had less pejorative connotations, meaning a man who worked the soil, who was simple, without artifice, and rough and dour because closer to nature.

Here, we are dealing with a quite different view. Instead of villainy and vile actions, the qualities of the rustic are those of someone who lived in a primitive and therefore archaic society. Sixteenth- and seventeenth-century poets, novelists, and playwrights perhaps abused the pastoral genre when they described societies of men and women, living close to nature, free from the cares of ordinary folk, in a sort of poetic "Golden Age." Yet both they and the moralists brought out the fact that "real" country people were, in truth, anterior, more archaic, and closer to the natural life than others.

There is no shortage of texts in which the rustic villager was compared to the pagan, the savage mentioned by travelers and explorers of newly discovered lands. These travelers (missionaries and soldiers) sometimes found evocative similarities between Indian customs and those of the villagers in their own country, and when they described rites and beliefs they resorted to a vocabulary commonly used for the description of superstitions, witchcraft and other practices rife amongst the common people in their own society. The villager – as in Antiquity – was characterized by two coherent tendencies: naturalism and pessimism. The former was an inevitable consequence of his work. The latter was already to be found in Antiquity as an expression of the belief that nature herself is cruel. However, the peasants' naturalism served as a basis for the ordering of work and festivals of a Christian nature. Moreover it led them to consider that there were not only sacred times, but also sacred places.

The year was divided into four seasons, with months bearing Latin names, which were in turn divided into weeks, with days bearing mostly pagan names. It started more or less with the Birth of Christ, at the winter solstice, and in the middle of the year fell the festival for the commemoration of John the Baptist, which coincided with the summer solstice. Saints' days ordered the seasonal jobs, the different rustic activities of the farmers and shepherds, the basic festivals coinciding with sowing, harvesting, and the movements of flocks, etc. From the religious point of view, sacred space was represented by hermitages, sanctuaries, monasteries, places of worship in general whose fame ranged from the local to the regional, national, and even international ambit. The peasant lived closer to all this than did other men. His relationship with nature, with the basis of human sustenance, meant that in the theory which pictured society as a body similar to the human body, he was placed in an inferior position, as the feet. However, like the feet, he was seen to be the support for the rest of the body, and like the feet he had to be looked after.

This was not all. there were various reasons to consider that there were honorable people amongst peasants. One was their simplicity and innocence, their distance from the corruption thought to be typical of city life, which yielded its particular refinements. Another was that the peasant was likely to have purer ancestry, to be more of an "old Christian" than many nobles and gentlemen who had, for economic reasons, married descendants of rich and powerful Jews, so that their offspring were not of pure blood and could not claim the "purity" and honor so important in this regard.

This confrontation between old Christian villeins and nobles of impure blood was portrayed in many popular plays of the time, as is well known. In what could be called more serious theatre there were also portrayals of the conflict between the aggressive concept of honor held by noble men of lineage or daring warriors and that of the villein, which was based on the principles of Christian morality. *El Alcade de Zalamea* of Calderón de la Barca could serve as an example of this conflict.

The peasant then, is not a being who can be characterized by his homogeneous features, because, if on the one hand rural society was typified in bestial, rustic, or grotesque images, on the other it was portrayed poetically and at times in the most gracious of terms. The image of "the honest peasant" and that of "the rich farmer" were far removed from those of the villein and the rustic.

A similar multiplicity of images can be found when the other "estates" are put under close examination.

Honor and money – the image of the merchant

In principle at least, the image of the businessman, trader, or merchant seems less complicated and contradictory in his relationship with Catholic morality. However, examining in depth the very rich documentation at our disposal today, we soon realise that we must take account of the great variety of tones and contrasts, and of the need to rid ourselves of assumptions culled from reading some famous works on the influence of ethics, in particular the protestant ethic, on the development of the modern capitalist economy.

We must indeed start from the idea that, until late in the modern era, all trading which entailed interest was, for the Christian, basically impious, and thus men dedicated to trading and making contracts could not, almost by definition, be good Christians. Jewish usury flourished under the influence of such ideas. However, as time went by, in the Catholic countries (and in Italy earlier than in Spain) moral theologians, confessors, and spiritual advisers had to develop new theories on interest, distinguishing between impious business, which would still be considered as usury, and pious business, brought about largely by what was called "novelty." This economic novelty was the result of the discovery of America, the strong trading links formed by sea with Asia, Africa, and countries far from Europe, the circulation of natural products, unknown until then, and the increase in manufacturing. Thus from early on the moral theologians of Spain and Italy are found holding forth on what might be called "economic casuistry" in such a way as to form two groups: one group was composed of the strict followers of rigid ancient tradition, while the other attempted to adjust themselves to "novelties" and innovated. We might call them "modernists." In any case, inspired by the conditions of commerce, they produced some curious metaphorical ideas, such as that of conceiving human life as a "spiritual voyage" in which one must contrive to arrive safe and sound in the port of eternal life or, to mention another, of estimating that this life is also a spiritual market place in which deals and contracts are made, some of them bad and some good. Furthermore, a strange quantitative criterion came to be applied to cases of conscience and sin in treatises of moral theology and in confessors' manuals, especially in the seventeenth century. The confessor not only had to qualify the gravity of the sin, but also find out the number of times it had been committed in this or any other form, in order to be able to quantify accordingly the penance given. The assets and liabilities were totted up as if in a commonday ledger. Such rationalization of the confession seemed monstrous to Protestants and free-thinkers, who thought it to be typical of the Jesuits. But the Casuists did not alone belong to the Society, and those known as "laxists" did not always have the majority.

However, one could not say that there was any theory constructed to give absolute, overall freedom to merchants, traders, or businessmen. Such people, especially those who had accumulated great wealth or had intricate types of business, frequently consulted moral theologians on the licitness or illicitness of an operation, and received complex, subtle replies. Nonetheless, the powerful merchants often ended their lives by leaving large bequests to the Church or founding pious institutions, such as hospitals, convents, churches, or chapels, which they endowed with fat subsidies. Many of Spain's religious monuments which have been thought to have been founded by the aristocracy were in fact the creation of rich businessmen of dubious repute, whose ill-fame was more often due to their "blood" (because they were thought to be descendants of converts) than to their professional morals. There were jokes about this: for example it is said that beneath the inscription on the façade of one such foundation, which read "El señor Don Juan de Robles con Caridad sin igual hizo este santo hospital" (Señor Don Juan de Robles of unequalled Charity made this blessed hospital), someone added "Y tambien hizo los pobres" (And he also made the poor).

Ever since ancient times a great distinction had been made between the small businessmen or poor traders, and the rich merchants with ample resources. As in the case of the peasant, of whom there existed one positive image and one negative one, so in the case of the businessman there were extremely varied and contradictory points of view. For while the image of the rich, ascetic businessman given in religious literature may recall some Protestant images, there are others of powerful men who steered a course between civil honors – they were able to obtain titles and enter chivalrous orders (although this went against the founding statutes) – and the suspicion of the Inquisition that they were possible Judaisers or related to people already prosecuted for Judaising. Especially during the reign of Felipe IV, there was often some smoothing over of the contradiction, found in extreme cases, that a family could at the same time be laden with wealth and honors and threatened or slandered. This was the subject of much legal argument over the fundamental concept of purity of blood, according to which the religious "guilt" of an ancestor could give rise to the incapacitation of his descendants, who, consequently, could not partake of public honors. The solution to these situations, often repeated, was – obviously the only thing to do – to falsify genealogies and bribe people to witness their validity. Satirical literature and the writings of the popular "moralists" of the day are full of allusions to the buying of titles and honors by people who were not worthy of them, either on account of their profession or of their origin, which rendered them theoretically "unfit." In practice it is known that in the decadence of the seventeenth century almost everything had a price: from titles of Marquis or Count to that of City, when a municipality which could afford it acquired the pretension to become one rather than to remain a mere town.

Soldiers

In general terms, as has been said, for the sixteenth- or seventeenth-century man given to speculation, the peasant was an archaic being. This point of view later gave way to the "pastoral" genre and Romanticism. Later still, in the modern era, it formed the basis of the theory, made famous by Spengler, that the farmer or peasant, "lacks History." In contrast, the businessman or merchant seemed to be the representative of modernity. And by modernity, it was understood that he also represented all its dangers.

This image was not in itself modern. Already in Antiquity, from the times of Plato and Cicero, there were various authors who contributed to its creation. The city, especially the maritime city, with its developments in the fields of technical, manufacturing, and artistic skills, represented that which was modern, and it entailed moral dangers as well, as opposed to the countryside, imbued with simple habits, customs, and virtues from the "good old days." However, within the society there existed other "Estates." Amongst these, two stand out as the most important, and they are related to each other. One is the old aristocracy. The other is that of the soldiers of the time. As in the previous cases, Catholic moral theologians and moralistic observers had a lot to say about them. The noble and the soldier represented, more than anyone else, "the man of honor." But these men of honor also figured, yielding to their vices, in the "Dances of Death" and in the confessors' manuals which accorded them special space on this account, for they were known to be prone to the specific sins of arrogance, violence, cruelty, etc.

Just as the conflict between theory and practice led the sixteenth- and seventeenth-century moral theologians to suggest new ideas about business, so, much earlier, the warlike organization of medieval society had led to the creation of a new concept of Christian notions. The Church was militant. The Christian was a soldier of Christ, the warrior against the infidel. The army of the faithful includes both soldiers and captains and a distinction can be made between the old legendary and popular captains and the modern, more effective ones. Among the legendary captains one might point to the Archangel Michael, lord of the mountains, patron of shepherds and fighters in times of resistance, and Saint James – not in his guise as the patron-saint of religious pilgrimages, but as the hero who won fabulous battles, Santiago Matamoros (the Moorkiller). In the eastern part of Spain another legendary warlike old captain was Saint George, known as Sant Jordi to the Catalans, whose collective patron-saint he was. The modern captains were the saints of the Counter-Reformation, headed by Saint Ignatius Loyola, who gave the Jesuits their seemingly military organization, as "the Company of Jesus." The faithful were also organized in a military manner, possibly more so in Spain than in other countries. This can be accounted for by various factors: on the one hand the power and influence of the long Reconquest against Islam, on the other, the imperial enterprise supported by the Church, which thus gave its support to an organisation of the State that was geared to combat and gave birth to a kind of patriotism both warlike and religious at the same time. This has been the basis for many commonplaces which have had repercussions that can still be felt today. Plays, biographies and sermons helped to popularize the warrior figures as being at the same time Christian and also national heroes.

To look at the other side of the coin, the moral theologian, the preacher, the casuist, and the confessor knew only too well that the daily life of the soldier, whatever his rank, could be, and generally was, profoundly immoral. The soldier was a robber, drunk, gambler, lecher, and blasphemer – and proud of it too. In the street he could boast of his sins, but in the confessional he would talk in a different manner. The moralist had to walk a tightrope for him. But in soldiers' autobiographies and memoirs, which are extremely illuminating on the subject, it can be seen that Christian morality and his code of honor, both moral and professional, coexisted and came into constant and dramatic conflict as soon as the soldier in question ceased to be a mere instrument of violence and acquired a minimum of conscience.

The aristocracy

Similar problems, though not the same ones, are to be encountered when dealing with the nobility, the aristocracy of pure race and blood. The ancient nobility, to be distinguished from that which was created with freshly made money, was founded on the same warlike tradition and on an ethic which was not originally rooted in Christianity. For the rules of conduct were dominated by the idea of pride in lineage, which was, furthermore, pride in race. As in the France of the ancien régime, where the aristocracy based their class privileges on their descent from the conquering Franks, and thus gave rise to a theoretical race history which had its repercussions down to the nineteenth century, so in Spain the aristocracy claimed to be descended from the Goths, i.e., another race of Germanic conquerors. From this comes the expression *hacerse los godos*, to "behave like the Goths," which was an allusion to excessive pride. According to the genealogists at the end of the Middle Ages, almost all the aristocratic families descended from these Gothic conquerors . . . even those who came from the parts of the country where the Goths had never held dominion.

This concept of "by dint of origin" was accompanied by a concept of life which could be considered Heraclitean in the sense that it accepted life as being essentially a battle. We should not try to look for metaphysical truths in this concept, since the people who subscribed to it believed that the battle was centered on the defense of the claim to greater worth (*mas valer*) of the claimant and his lineage within a closed society. Chroniclers of the quarrels and wars between medieval clans and families in the Basque country and its surroundings, e.g., Lope García de Salazar, gave a clear portrayal of this thesis. "Greater worth" was imposed and conserved by use of arms. The fight for small honors and precedence was continual and horrific. However, roughly a century later, when society was already somewhat changed, some men with old-fashioned ideas like Lope de Aguirre took the concept with them from this region to America, as a part of their imperial enterprise. The fights between factions and lineages led by their chiefs and family-heads produced an anarchic situation which ended in the strengthening of the monarchy. The aristocracy champed at the bit, and some of them, openly hostile to the order imposed, looked for support to a bourgeoisie of bureaucrats and businessmen, and even to a popular element. The nobleman, nonetheless, continued to be an important person, and the Church recognized this fact in a highly elaborate, complicated system. While it had to proclaim the equality of all men before God, and defended the supremacy of spiritual power, it recognized, at the same time, the existence of the superior temporal power of the emperors and Christian (Catholic) kings, and with it the privileges of the nobility. Whatever the case, the representatives of spiritual power as much as temporal power and those who were privileged by birth were all sinners. Confessors dealt at length with subjects such as the most frequent sins of Princes of the Church, lords of vassals, nobles, and men of lineage, and their arguments about these sins caused rifts and polemics. Related to one of the central themes of this essay were the discussions of the sins committed in defense of honor.

Much has been written on the subject of this kind of honor, not all of it good. In the noble it was something similar to the Greek hubris, a violent disposition that produced constant provocations and insults as a result, and challenges which frequently ended in a death, especially among the young. How could such a situation be reconciled with evangelical gentleness? Some theologians, the rigorists, stated clearly that there was no possible reconciliation. Moreover, there was no shortage of lay moralists, including the aristocrats themselves, who wrote against dueling and protested that honor was being confused with what they called "point of honor," from which ensued challenges and violence. However, there were several casuists amongst those called laxists or partisans of the "lax morality" who supported the argument that one could go to extremes in defense of honor without committing a grave sin. The representatives of this "lax morality" which, through excessive generalization was also called "Jesuit morality," saw clearly that it was one thing to live as "a good Christian" and quite another to live in "reality as it really is."

Conclusions

However, this position was considered unacceptable by many and was almost totally discredited after the memorable struggle between the Jansenists and the Jesuits, whom men such as Pascal in France condemned as representative of the most depraved laxity. It must be recognized first of all that the examples used by Pascal in *Les Provinciales* were selected from amongst hundreds to produce the desired effect. The casuist texts considered to be the most scandalous, such as Escobar, gave an overall impression that they were not in fact quite so acquiescent as was suggested, and that their conception of the human conscience was frightening: they may have been nearer the truth than other authors hostile to them, when they accused the Catholic moral theologians of taking delight in the morbosity, in imagining all sorts of abominations and aberrations. The books by casuists deserve to be the object not only of historical study in the present day, but also of psychological and psychoanalytical research. On the other hand, the quarrels in the ranks of orthodox Catholics were considerable: each was against the next, Dominicans, Jesuits, etc., and those who, at the end, were thought to be the most lax were condemned.

All in all, society readjusted itself. The "Gothic ideals" were already the object of general derision by the eighteenth century. If in the thieves' cant of Cervantes' day[3] "Gothic" meant illustrious or noble and "Goth" rich or eminent, for the writers in the eighteenth century these words were synonymous with archaic, and therefore by the same token extravagant. However, until the Cortes de Cadiz in 1812, it could be said that society divided into "estates" remained in force and that the characteristics and contradictions herein described continued to exist. Nowadays there is a tendency to forget this.

Notes

1 *Las Formas Complejas de la Vida Religiosa. Religion, Sociedad y caràcter en la España de los siglos XVI y XVII*, Madrid, 1978.

2 Translator's note: The Spanish word *villanía* can mean both "villeinage" and "villainy." Similarly, the word *villano* can mean both "villein" and "villain".

3 Translator's note: The reference is to *Rinconete y Cortadillo.*

6

The *sophron* – a secular saint? Wisdom and the wise in a Cypriot community[1]

J. G. PERISTIANY

I am concerned in this paper with wisdom and the wise and with their relation to honesty, honor, sin, leadership, and the ideal image of man in the highlands of Pitsilia.[2] This, in reality, is an attempt to see social rules and the evaluation of behavior from the perspective of the village community – a personification that renders meaningful and effective what may otherwise have been seen as an anonymous social order. This is one of the main tasks of the *sophron*.

In pursuing this analysis I chronicle at some length both local events and the reactions they caused in the hope that this may ground Pitsilia value judgments in the villagers' own perception of social reality.

A reference first to accepted usages in dating village history and then to my first encounter with a wise man may serve both to introduce the subject and to point to its significance within the village context.

Pitsilia's relatively distant past is often dated by reference to events, or to persons, that have set their seal on village history. These events may have a purely local significance. They may, also, be local reflections of a wider phenomenon. When seen separately from what they have come to signify they often appear trivial. Thus, in Alona, the purchase and installation of the first (iron) bedstead marks, to this day, the end of World War I and the moral disintegration of the community. Reference to the arrival of the first *karkola* dates, according to the context of the conversation, a contemporary event or marks the moral character of a period.[3]

The description of the arrival in Alona in 1927 of the first motor vehicle, of its color and make, of the effect of the car's backfire on women and children and of its final plunge into the ravine marks, to this day, the effective opening of the road to Nicosia which reduced the day-long mule and camel-train journey to under four hours. It also refers to the urbanization/satellitization of the village.

Again, the desecration of a local ikon is used as the local reference to the hanging, in 1821, of the Archbishop and the notables of Cyprus and the rise of national and religious intolerance. And the last feasting of Turkish policemen on village hens, in 1878, is related to the arrival of British troops and the resulting freedom from highhandedness.

In these examples the refraction of the general through the particular provides the general with a local, an esoteric, meaning. It integrates in it the imagery, symbolism, and speech which mark the exclusiveness of village society. It thus serves as a signpost of local history and as an element of local identity.

Dominant local personalities impress themselves upon, and are often identified with, longer periods of local history than those associated with mere events. The period of ascendancy of those whom public opinion recognizes as communal leaders distinguished by their wisdom is referred to by their name, e.g., "in the days of Hajjisava," a period which may stretch back over a hundred years. Until the early 1960s there was practically, with the caveat I shall add later, no interval of time when such a connection could not be established. Thus, a rough outline of the history of Alona may be dated by reference to a succession of wise leaders, their names and their deeds being committed to memory, so that they become paradigms for the young, both as singular historical personages and as notable exponents of accepted rules of behavior.

I have indicated the dating function of the local refraction of phenomena associated with a wider society. When this is the case, their uniqueness, their non-repetitiveness, associate them, logically, with a lineal concept of time. This applies equally to the names and to the idiosyncracies of wise men, as well as to the forms taken by their sagacity. Historical lineality applies less accurately to the words of wisdom and to the precedents with which they are associated when these are used to buttress the wise man's advice.

Wisdom itself is impersonal. It forms part of a common fund, a crucible, with robust assimilative powers so that everything entering its mainstream is made to appear as compatible with an ideal pattern. Quotations from classical Greek writers, from Greek history from Homer to the present day, from the New Testament and hagiographies to the sayings of notables feed the pool of *sophia ton progonon* (the wisdom [always in the singular], of [our] forefathers). The precepts of wisdom, as long as they are believed to be integral parts of the Greek or the orthodox heritage (the two being conceived as coterminous), are perceived as complementary views of a harmonious and perennial whole whatever the apparent contradictions. Thus, to strive always to excel, "to be the best and to stand above [all] others,"[4] is not seen as incompatible with the ideal of Christian meekness. They are both "proper advice," as long as they are directed to different moments of social or individual existence.

The conciliation of apparent contradictions, necessary in order to preserve faith in a homogeneous national identity, falls within the domains of teacher and priest. The application of these same principles to particular cases is the task of the *sophron*. His art is revealed in the adjustment of conflicting interests and in the defense of the harmony and reputation of his community. The most usual characteristic by which the last[5] *sophron* of Alona is remembered – "etan anthropos pou agapouse ta koina," "he was a man who cherished the commonweal" – is a phrase used again and again to describe this singular propensity.

One of the main criteria of the *sophron*'s wisdom consists in applying general principles to particular cases in such a manner that the general is not seen to bend so as to serve practical ends. Manifest casuistry is disparaged by Aloneftes not only because it undermines the credibility of their beliefs but also through its association with the "Jesuitism," of the "Franks" who are accused of plundering their churches and reducing them to servility in the name of Christ. "Jesuitism" was also, for Aloneftes, the main moral characteristic of the colonial government. Thus, what distinguishes the wise from the merely knowledgable is that the wise man knows not only the rules but also how to apply them for the common good. Mere knowledge pertains to "professors," a most derogatory term used for those who in the words or our last *sophron* "talk wisdom as though it were a dead language." Knowledge joined to the capacity to apply it is a constituent of wisdom. Pure knowledge is an ingredient of foolishness.

The *sophron* uses received wisdom more than he contributes to it. In this sense wisdom is static and extraneous to a particular village community. The *sophron* cannot act as an innovator for the same reasons that he should not allow himself to be cast in the role of casuist. The *sophron* can neither order nor decide; he can only advise. For his advice to be followed he himself must appear to be the servant of the community and the mouthpiece of wisdom. As for wisdom itself it commands assent as long as the assumption of its timelessness remains unquestioned.

My first encounter with a "wise man," the *sophron* of Alona, took place on the night of Good Friday, 1954. I was exploring the highland villages of Troodos, an anthropologist in search of a model, when I reached Alona with my fellow driver, Niazi-the-Turk, as dusk was falling. The houses and the streets were deserted, the coffee shop[6] shut and there seemed to be no refuge from the cold and the damp diffused by low-flying clouds. Niazi ruminated over our having to spend the night in this "haunt of ghosts" but we had no heart for traveling by night on the narrow and precipitous road that led to the neighboring village. We followed through the hazelnut bushes the sound of a bell tolling until we came to a church plunged, but for the flicker of candles, in almost total darkness. The men on one side, the women on the other, all dressed in their darkest clothes, were kneeling and chanting a dirge. The scent of incense and of burning beeswax mingled with that of massed humanity.

Through elective affinity and a quickening of the senses an immediate communion was established with the passionate fervor of the congregation.

I asked the priest, after the service, whether objections might be raised to a Muslim spending, in the village and with a Greek orthodox family, the remaining days of Holy Week. The priest said that, as we are all children of God, Niazi would be welcome at his church but that he would prefer Hajjiyannis, "a man with a well-attuned brain," to answer my question and to persuade an eventual host to provide him with shelter.

Hajjiyannis was a tall, lean, ascetic-looking man in his late sixties. He had the spiritually commanding presence of a man before whom other men fall silent. My task was not easy. I had to explain how I, a Greek Orthodox, was visiting during Holy Week one of the most devout villages of Cyprus with a Turk – a necessity for my work in mixed communities. I went on to add that attending religious services not only in our own church but in synagogues and mosques formed part of a family tradition. Hajjiyannis approved, in private, what his fellow villagers would certainly have considered an eccentricity.

I came to know Hajjiyannis well. I was attracted to Alona and its people. An enduring bond which continues to this day was established that Friday night in church.

When I came to live here Hajjiyannis explained to me how he had proceeded to solve my problem. He had called upon some of the leaders of public opinion, the priest, the teacher, the *archontes*,[7] and pointed out that the friendly reception of strangers was an ancient Greek, as well as a Christian, virtue. To refuse hospitality to a Turk, especially during Holy Week, would have discredited Alona. But how to impose a Muslim on a family – and which family – during the special devotions of these holy days? Hajjiyannis called, and that was the first time he had done so, on a newly enriched but socially marginal family of recently returned emigrants, originating from shepherd stock. You have vouchsafed much honor upon us, said the head of the family, when asked to act as host on behalf of the community.

This appears to be – indeed it is – a simple story: the manifestation of a person who, without holding any office whatsoever, is endowed with sufficient moral authority both to reconcile conflicting values and, if asked to do so, to speak in the name of the best interests of the community. These conflicts may actually appear more acute, and the role of Hajjiyannis more arduous, when we consider not only the tensions of Holy Week but also the fact that, according to local belief, a Turk had attempted to desecrate the miraculous ikon of Alona ("when his right arm withered through divine retribution"), that no Turk lived in these mountain fastnesses and that this was the eve of the Cypriot struggle for independence in which Greek Cypriots opposed what appeared to be a British–"Turkish-Cypriot" alignment. Only a temerarious – and insensitive – stranger would have insisted on overruling public opinion at such a moment, had this not been prepared by the thoughtful actions of Hajjiyannis who, as I came to learn, was not only a "sensible" person but the *sophron* of Alona, albeit at times reluctantly. He became my friend and, in matters concerned with his social role, I was his confidant. This account is, of necessity, influenced by this relationship.

These actual examples of the work of the *sophron*, which now form an integral part of village history, will situate my theme.

The alien groom and the threat of "anomie"

The marriage around which centered the conflict that Hajjiyannis was asked to resolve was already marked, before its celebration, with the portents of failure.

Two of the main ingredients of a successful Alona marriage are village endogamy for those who reside in the village, and isogamy or hypergamy for the bride. In the first instance, that of village endogamy, my latest (1983) work in Alona shows that the actual is closely modeled on the ideal in relation to the provenance of bride and groom living in Alona. My figures concern Alona men and women born between 1933 and 1963. Out of thirty-four Alona brides who had their home in Alona, thirty-two were married to fellow-villagers and only two had "imported" grooms from a neighboring village. Of twenty-five grooms who were actually living in Alona[8] with their wives, twenty-three had married Alona brides and only two had brought non-Alona wives to live in their village. Thus, in relation to this twenty- to fifty-year-old group, only two husbands and two wives were non-Aloneftes, living in Alona.[9] It is more difficult to assess statistically the elements of isogamy but it is unusual for a woman to marry conspicuously beneath her unless age, or social and physical stigmas both of which are often considered hereditary, leave open no other avenue to marriage. The physical intimacies which, even in proper wedlock, tarnish the image of ideal feminine purity are then seen as acts of defilement.

In this area uxorilocality results partly in female hypogamy as the man who moves to the bride's village, the *sogambros* (literally the live-in-groom), is treated as an inferior. When his status in his village of origin is inferior to that of the bride in her own, this compounds the woman's degradation. The expression used when such a situation is discussed, "our own maiden is defiled [literally is being soiled] in [her own] village," is indicative of the sentiments aroused by this situation. To my knowledge no marriage has proved successful during the past thirty years (this being the period of my personal experience of this area) when the imported groom was forced to reside permanently in the bride's village. In this case his status is, at best, ambiguous. To his many disabilities is often added the malignity of his new fellow-villagers, canalized by his in-laws who may have come to regret their share in their kinswoman's loss of dignity. Indeed, the *sogambros* is encompassed by an aura of mistrust. Cast, from the moment he takes up residence in his new village, in the role of "stranger/potential-enemy in our midst" the expectation is that one day he will betray its interests and its values as "what can you expect of a hybrid?"[10] The imported groom himself fears that his every word and action will be misinterpreted so that one day not only his affines but his own wife will side with their fellow-villagers and turn against him. With a *sogambros*, village economic and social cooperation is minimal. One of the three resident *sogambroi* eventually founded his own coffee shop whose only customers, during its short existence, were the other *sogambroi*. During the course of an anthropological conference I invited my colleagues to visit the village and have lunch at this coffee house. The invitation was extended to all my friends in the village. Only one other *sogambros* and the affines of the *sogambros*-owner came.

When the bride is "imported" she is usually referred to as "the foreigner," *e xene*, a telling and accurate description of her position in her husband's family and village. An alien she will remain to the end of her days. In her case it is presumed that, like all *xenes* (that is all women who did not grow up under the vigilant and inquisitorial observation of their fellow-villagers) she will, at some stage in her life, give way to her latent immorality thus betraying her husband and the good name of the village.

The bride, in the case that I shall be discussing, was young and fair. Although endowed with the additional attraction of being a brotherless heiress, her domineering and arrogant nature repeled suitors from her own village. The ageing parents, who needed a young man in the house and in their fields, arranged for a *sogambros* to marry their daughter. A poor boy from a neighboring village accepted this role. He came to live with the family of his bride-to-be and, by becoming his in-laws' drudge, eventually lost his good name and forfeited his manliness. The string of nicknames lengthened with the increasing opprobrium in which he was held. The "adventurer," *tychodioctes*, the "carrier-of-burdens," *hamales* (here a Turkish word was used to aggravate the insult), the "goat-herder" (a feminine role), etc. His own betrothed, who had lost all respect for him, treated him like a menial and, as the witnesses to his degradation put it, "what had to happen, happened." In this case his challenged masculinity asserted itself by making another girl pregnant.

Village opinion was faced with two conflicting and seemingly irreconcilable sets of interests. The future parents-in-law wished the man they had introduced into the community to leave the village. His presence embarrassed their daughter and acted as an apotropeic to eventual suitors as it served as a reminder of her relative unworthiness, the physical proximity of a man presumed to have been intimate with a woman – in this case during the period of engagement – being said to shame and emasculate a later husband or lover. The parents of the girl he got with child wished him, with equal vehemence, to remain in the village and marry their daughter so as to repair the damage he had done. Time was of the essence of the problem as kinsmen and friends of each family were rallying to its cause and the pregnancy was advancing to its natural term. The public opinion of the in-groom's village also had to be considered. Their man had been hiding to avoid assault and his humiliation, deeply felt and unavenged, reflected unfavorably on their own village.

It seemed, at this stage, that the matter could not be settled without recourse to some form of violence attended by the sequels of internal schism and external strife. The exhortations of priest and *archontes* were of no avail in this contingency, the priest's latitude of action being qualified by his calling while the credibility of prominent villagers was limited by private considerations, mainly those resulting from the primacy of their bonds to either or both the conflicting families. Neither priest nor *archontes* could initiate a course of action which would result in both lovers leaving the village, the radical solution advocated by Hajjiyannis. This, he explained, would provide all concerned with a peaceful interlude during which things might sort themselves out. Indeed Hajjiyannis, although closely connected to one of the families, had been consulted by all parties – a clear indication of their belief that his wisdom transcended his personal loyalties. His advice never varied. Since antiquity exile has been considered preferable to schism.

The lovers themselves wished to marry but they were reluctant to leave the village. Neither could be forced to do so unless the very marriage they were contemplating could be made to appear, in the eyes of the bride's fellow-villagers, as equally shameful with their present condition. A charivari assuming the form of a *shyllogamos* (literally a "dog marriage," here the coupling of dog and bitch) was organized. This was performed outside the home of each lover and was intended as a re-enactment and a debasement of their prenuptial sexual union. The dog, so close to man, may be used as an illustration of man's virtues and of his frailties. As a socialized animal the dog exemplifies the approved characteristics of selflessness and loyalty-unto-death while when its unredeemed animal nature is permitted to take over its actions become paradigms of what would happen to man when, undeterred by shame and guilt, he submits to covetousness, gluttony and, above all, to unlawful – and therefore bestial – sexual desire. In this parody the "dog," led by its instincts, scents the "bitch" and mounts her publicly. Having done so his "nature" assumes control over his actions and he cannot uncouple until it has taken its course. The locking of dog to bitch, the mechanical and almost irreversible nature of their action once the animals have joined, is the subject of the charivari. Young men, shouting abuse and beating upon pots and pans, dance round those simulating the union, trying to shame the performers[11] into uncoupling, while they, clinging to each other, cry out that they are unable to free themselves. The voluptuousness of the devil is expressed through canine orgasmic behavior. This simulated copulation which neither of them can interrupt or undo is said to express the lawless and uncontrollable lust through which the devil and animality triumph.

The *shyllogamos* had not been performed since the end of the last century. It is so shaming – and this was the purpose of its present performance – that the two lovers could not have gone on living in the village as their joint presence would have evoked the memory of their public disgrace. Following the charivari the *sogambros* returned to his home. His mistress joined him there in the hope of being married to him but his kinsmen and friends shunned her, for now "shame was part of her." Rejected, the expectant mother returned to her own parents. Here, pity for her condition took the place of resentment for her fall. Her kinsmen, seconded by the priest, arranged for her to be married to an elderly, childless widower, a fellow-villager who adopted the son she bore. Her return and acceptance, both of which were warmly supported by the priest and *sophron*, served as living illustrations of Christian charity while she became an exemplar, a warning, to village maidens, of the consequences that may attend the yielding to sexual temptation.

Matching the ideal to the practical is the essence of wisdom. When this process is reversed, primacy being assigned to principles while the actual problem is stretched on a procrustean bed of ideals, this is seen as non-pragmatic foolishness. The good doctor, say the Aloneftes (and the *sophron* is a doctor for social ills), uses as his starting point a particular patient's illness and then determines the cure.

The *sophron* and the ambiguity of the reputation of the village

The Pitsilloi are deeply conscious of their cultural roots, of what they call their *riziko*, their destiny, their *efthene*, their responsibility, to hand down to their children the ideal image of their identity, a blend of local custom, eastern orthodoxy and Cypriot Hellenism seen by them as pure and unalloyed.

The people of Alona and their highland neighbors are amongst the most ardent in the fields of knowledge and warfare. Alona, already a *prestrerie*[12] in the thirteenth century, has provided its neighbors during the past hundred years with eight Alona-born priests – a source of immense pride and a means of influencing its neighbors. Soldiers and guerillas have also come from this area. The Alona vision of Pitsilia, a paradigm for Cyprus, and of "Alona, an example to Pitsilia" is instilled in their children. In my survey of over 25,000 schoolchildren[13] of western, central, eastern, and southern Cyprus, the pupils of the primary school of Alona have scored some of the highest average marks of any Cypriot rural area as well as one of the highest percentages in the island of follow-through to secondary, technical, and university studies.

Beside their fame for bravery, patriotism, and the pursuit of knowledge, which contributes to both their individual and their collective reputation, the people of this area are noted for their thrift – in this case a mixture of frugality, of hard work, of avarice, and of rapacity. It is the notoriety of their defects, not their cure, that calls for the mediation of the wise man so that the good name of his village may be restored at a time when the extortionate actions of its inhabitants tend to bring it into disrepute.

This is Theodoulos, painter and recorder of village history, speaking:

> In Alona, the forces of good and the forces of evil are forever contending not only for the soul of man but also for the good name of the village. It is a wondrous thing that their extremes should cohabit here as husband and wife. We men of Alona spend our lives endeavoring to reconcile them. This is a hopeless but necessary task as the forces of goodness bestir themselves when evil threatens to overwhelm us. We assist them with the help of God and the advice of those to whom he has given *sophrosyne*.

The use of enigmatic and sententious expressions, especially when they assume the form of well-known aphorisms or parables, lends weight to the remarks of those who lay claim to wisdom. The use of a similar verbal mould for both the original saying and its present derivative reinforces the belief in their common origin. An added advantage of this form of speech is that it not only draws the listener into the conversation but allows the proponent of wisdom to use him as an associate, as an involuntary "compere." Parables and aphorisms are commented upon and examples from actual life are provided by a cooperative audience. Those are then woven by the *sophron* into his original sayings so as to extend their use. To simplify these sayings by metaphrasing them, by making their meaning immediately accessible, would be churlish as this could but limit the range of their use when their function is to serve as master keys; the more hidden the "truths" they help to reveal, the greater their versatility. A wise man uses his audience so as both to strengthen his sayings with *peitho*, persuasiveness, and to enrich their hermeneutic value.

Image-building. The "sophron" intervenes

One of the recurrent problems of Alona has been the changing reflections on its reputation of the contending forces of good and evil. While often praised for the achievements of its inhabitants the collective image was at times tarnished by the sharpness of their commercial transactions. This both shamed them and reflected adversely on their dealings with their neighbors and their far-flung clients. Thus, both ideological and practical reasons placed the problem of their image – "what are other people [literally the world at large] saying about us?" – at the center of their preoccupations. The two examples I am recording here point to the importance of some of the problems submitted to the *sophron* as well as to the moral ascendancy which helps convert his advice into action.

The men of Alona travel throughout Cyprus peddling the produce of their labor: smoked ham,bacon, wine, hazelnuts, and almonds. Frequently these are sold not on the open market but directly to urban patrons. This marketing technique eliminates middlemen and establishes a network of patrons–clients who often share in common ventures, usually the purchase for ready cash – and the subsequent resale on credit – of donkeys, mules, and flocks of sheep. This mode of trading is, in reality, another aspect of their money-lending activities.

The intimate knowledge both of the areas they visit and of the financial standing of their inhabitants helps them to make quick profits which grow even larger when the borrower cannot repay his debt since then the creditor acquires the land mortgaged to him for a fraction of its value. The "thrifty" Aloneftes have thus come to own substantial plots of land in neighboring villages. "These fields," said Theodoulos, "are the rewards of *krimata*, of sinful behavior. These *krimata* are 'bitter' and Aloneftes find that they weigh heavily not on their honor but on their conscience. To lighten their burden they journey for confession to distant[14] monasteries whose monks impose heavy penances upon them." But the relief of individual consciences does little to restore the reputation of the village which, in history past and recent, was frequently associated with money-lending and extortion rather than with the ideal image of their community which its inhabitants wished to project. The Aloneftes' claim that their neighbors' accusations were *avanies* (an *avania* is a false rumor deliberately put into circulation, a calumny) produced the crushing rejoinder that the *avania* was, actually, an Alona stock-in-trade. The play is on the word *avania* which connotes an extortionate tax once levied by the Turks.

It so happened that, at a time when the reputation of Aloneftes, accused of cupidity, had been deteriorating, the monastery of Makhairas[15] (to whose monks they often resorted for the confession of their sins) burnt down. Wise man Hajjisavva suggested that, rather than contribute individual donations, credit would be reflected on the village as a whole if Aloneftes organized shifts to work *pro gratia* on the monastery site. This was not easy advice to follow as it disrupted the cycle of both their trading and their agricultural activities and left the field of finance open, for long stretches of time, to their competitors. Among the many sayings used by Hajjisavva to support his advice were examples from the naval battle of Salamis down to the Greek War of Independence, supporting the view that in times of crisis a community should act as one body and that the interests of each individual could not, in the final analysis, but profit from sacrifices made for the common cause. He supported this view by tagging at the end of his advice his own pithy saying to the effect that the reward of the volunteer builders would "sprout" from their own labors. The priest and the teacher strongly supported this scheme which had been viewed with some misgiving by some wealthy and well-born leaders, the *archontes*, who were amongst the main village financiers. In the end they, also, were swayed by public opinion – strongly in favor of Hajjisavva's scheme – although it seems temerarious to accept what has now become part of Alona history, that Hajjisavva had, at the time when he was giving this advice, predicted and explained the material consequences of the action he was advocating.

A large number of Aloneftes worked in relay on the rebuilding of the monastery. To-day some of the restored columns are inscribed with the names of their individual artisans, followed by the word "Aloneftes," man of Alona. As a result of this action Alona gained collective renown among its neighbors as "the village of the benefactors of Makhairas." Notice was also taken of the quality of the workmanship of its inhabitants. This dual result reflected, in turn, credit on the wisdom of the *sophron* who advised this course of action. Today it is cited as the consummation of *sophrosyne*.

The second example of the restoration of village reputation by the *sophron* is situated in the period preceding the First World War. A neighboring village ran out of money while building its new church. It was decided to ask Alona, "The village of money-lenders," for a loan. A delegation was then dispatched to Alona to borrow £200 over a period of three years. This was a considerable sum at a time when a shilling-a-day was the wage of a laborer. The Alona *archontes* met. They collected the money and placed the 200 gold sovereigns in equal piles on a table behind which sat Hajjinikolas, the wise man on whose advice they were acting. When the borrowers arrived they were feasted by their hosts and Hajjinikolas handed over the money to them but refused to discuss the rate of interest. He also refused to accept a receipt as though the loan were being made to a spiritual kinsman.

Three years later the borrowers returned and ceremonially placed 200 gold sovereigns in piles of ten on a table behind which sat Hajjinikolas, the whole procedure being an inversion of the original meeting. When this was done they formed seven separate piles with an additional seventy gold sovereigns – which they considered as reasonable compound interest for the period of the loan. This was the great moment that the Aloneftes had been waiting for. Hajjinikolas asked the borrowers to take back the £70 interest "as we lend money and ask for interest because life is hard and we must take care of our families – but we do not accept interest from the All Holy (Virgin)." Hajjinikolas then placed a bag containing an additional twenty sovereigns, as the contribution of Alona to the completion of their neighbors' church, next to the seventy gold sovereigns, the interest he had refused to accept. Whenever, nowadays, the village is blamed for usury this redeeming instance of its munificence is retailed.

I have chosen these examples of the sagacity of the *sophron* as they form part of the hallowed lore of Alona. In this context, they are representative of what is considered as wisdom, a course of action coolly planned in advance in order to achieve results beneficial, in the first instance, to the community as a whole while using and consolidating communal belief in shared ideals.

From judiciousness to *sophrosyne*, from prudent man to *sophron*

Prudence and common sense are not the preserve of the wise. Like most social characteristics they are shared, although unequally, between other members of society. Family matters are discussed with kinsmen, important personal problems with one's sole "best friend," the *bisticos*, the trusty, impersonal matters with the *parea*, the group of – usually coeval – friends, public affairs by all. In these exchanges the opinion of certain persons attracts more attention, is given more weight, than that of others. Apart from the *sophos*, the learned, to the ambiguity of whose position I have already referred, a discerning man of sound judgment may be said to be expressed by *nousimos* or *myalomenos*, both meaning sensible, or *phronimos*, prudent. Although these terms may be used interchangeably, the first two stress awareness while *phronimos* denotes the expression of this awareness in actual judicious behavior. The persons endowed with these qualities are not devoid of *sophrosyne* which, apart from other faculties, encompasses perception of what is right as well as prudence in action. It is when we move from wisdom to the wise, from the quality to the person embodying it, that we find that the former is more extensively applied than the latter. Thus a village may be inhabited by a number of sensible and prudent men endowed with a measure of *sophrosyne* but it only admits to one *sophron* for he is not only the main modeler but the unquestioned mouthpiece of public opinion. As each man may have only one *bisticos*, one best friend, it is said of the *sophron* that he is the *bisticos* of the village. A rival would have undermined his uniqueness as catalyst, which is the main purpose of his social existence. It may also come, as in the case of Alona since the death of Hajjiyannis over twenty years ago, that no undisputed *sophron* is acknowledged. The effects of this hiatus reveal, as we shall see, the significance of his social role.

A man does not suddenly emerge as a *sophron*, neither is he groomed for, or appointed to, this dignity. *Sophrosyne* is gradually revealed in speech and action consistent with the image of a selfless and judicious man who applies these qualities to the welfare of his community. The *sophron* – and I regard this as one of his main characteristics – is a dedicated man who should allow no other aspirations, private or public, to distract him from his main duty. It is only the undisputed recognition of these qualities by his fellow-villagers that warrants the use of the term *sophron*.

I shall turn first to the conditions of sex, age, social rank, and character necessary to the acceptance of a man as *sophron*, and then examine the relations between the *sophron* and the system of values within which he operates.

Sex

Most virtues – honesty, honor, *sophrosyne* – assume their full, their widest, meaning only when they refer to masculine actions. When applied to women, their immediate reference is usually limited to a qualification of their modesty, or to their behavior towards kin. The very opposite is true of men. Not only do a variety of criteria derived from private and public activities enter into the evaluation of their social characteristics but prolonged sexual diffidence tends to undermine the efficacy of (I believe, all) their other virtues. Moreover, no woman in this area has ever been called *sophron*. Even *sophrosyne* seems to be beyond her reach. Here *phronimada*, prudence, sets the limits to the highest achievement of a truly feminine woman. Beyond this point lurks the danger of her masculinization – or the contingency of her saintliness.

The first requirement regarding the *sophron* concerns his masculinity. This, for all notables, should not be in doubt as a deficiency in this area would vitiate all other qualities. From the time of his birth to that of his marriage a Pitsilia father takes pride in his son's manliness. When a son is born to him he boasts of the infant's sexual conformation. A monorchid and a poorly endowed boy are ranged in a marginal sexual category. Their manliness is constantly reviewed and the resulting insecurity renders them unfit to become social leaders. It is praiseworthy for a male teenager to remain a virgin until the end of his secondary studies. This is an anxious time for the father, indeed for both parents, who are relieved to hear of their son's first sexual experience.[16] From this moment – until his marriage in his early twenties – the father considers his son's sexual activities as a projection of his own masculinity, especially when these take place, as they should, in an urban environment so that village propriety is safeguarded. This is, also, the only time when (on being questioned concerning his younger kinsman) a man may answer *eginen latzinarki*, he has become a stallion, or *kaprefki*, he is behaving like a (wild) boar, to denote affectionate pride rather than censure. Belief in sexual experience as a constituent of masculinity is so ingrained that men lacking it are regarded by both sexes as *nepia*, infants, and *agouroi*, unripe (like fruit not ready to be eaten), or worse still by two expressions used mainly by females to manifest their utter contempt for a sexually untried youth, *myxiariko*, a snotty brat, or *agamos*, here in the literal sense of a man who has not known a woman. A man ridiculed by women cannot command the respect of men, so that, together with physical disability and impotence, this shortcoming constitutes an impediment for a potential *sophron*. To have been *archatos*, heavy-testicled, as a young man is, for the community, a warranty rather than a hazard for it serves as an insurance against the prevalence in later life of faintheartedness and especially of petty malice which are said to mark the actions of women and of sexually inadequate men.

Age

No man is said to attain *sophrosyne* before reaching old age – here situated in the early sixties. Age is seen as a calamity and a haven. It is a calamity as, apart from the natural ills that follow in its wake, it forces Aloneftes to distribute their fragmented land among their offspring so that, as the parents come to depend increasingly on their children, a role reversal gradually takes place. It is a haven, as the intense competition for honor of early youth which is followed by the acquisitiveness of married men, struggling to provide for their families, to educate their children and to dower their daughters, is replaced by withdrawal from strife, eventually a total withdrawal, in keeping with their final duty which is to make their peace with God and their fellow men.

Thus the *sophron* should retire (as Hajjiyannis had frequently attempted to do) before the final stage is reached as involvement in social strife causes old men to relive their past, to go over the terrain of old battles and to fantasize how their lechery might have been satisfied if only they had asserted their manliness more forcefully.

Character and social condition

When I was retracing the life-events of the men in whom Alona took pride, it became clear that those connected with wise men, and in particular with the last *sophron*, a layman and not a priest, were those which were most prized. The features of his behavior which were singled out as being characteristic of the man were also those which were used to construct the image of its ideal representative.

Hajjiyannis, apart from his distinctive masculinity as a young man and his proven qualities as a *sophron*, could not be faulted in his behavior as a son, brother, husband, father, as schoolboy, farmer or elder. His everyday actions, as narrated, were those of a sensible average man. Indeed, exceptional achievement in a specialized field unconnected with his later vocation may have been seen as detracting from his fitness to act as a man who advises and speaks on behalf of the community. Alona has bred men with exceptional talents in many fields – technology, the arts, and banking – but none has been called a *sophron* as it is said that their specialized knowledge of one field detracts from their understanding of the others, of the blend of farming and trading which were the common pursuits. Fears are also expressed that their advice may be swayed by sectional interests.

An apparent contradiction may be noted here. I am referring to the junction in the same person of the office of priest and of the role of *sophron*, a not uncommon combination until the end of the nineteenth century, when the priest was still considered a universal man, a master of all forms of knowledge and wisdom. This is rarely the case at present when he is increasingly seen as a specialist identified with his cloth.

Before the founding of the first school, *c.* 1886, the priest was, invariably, the local, rather rustic, schoolteacher. Boys who wished to learn how to write and read followed him as willing helpers, when he went out to tend his vines and fruit trees, and recited the formularies of the Liturgy while they worked. They wrote and read when resting. The most proficient became themselves priests and, in a remarkable number of cases, their descendants followed this tradition. Thus Alona at the end of the last century was not only able to supply priests to neighboring villages (it was known as *papadochori*, the village of priests), but boasted of seven resident autochthonous priests, one of whom was the village *sophron*.

The difficulty for the priest to combine his functions with those of a contemporary *sophron* is not due to withdrawal from daily life. He is a married man, usually with numerous progeny, a witness both to his virility – with which he is traditionally associated – and to his abstention from any form of birth control. The priest is also a farmer and, as an administrator of church property, a businessman of keen discernment and a man active in the field of politics as an advocate of the "national cause." Neither could the present priest, nor most of his predecessors, be said to have fallen short of *sophrosyne*. What the priest lacks, in the eyes of the present generation, is the lay *sophron*'s independence. Being the subordinate of the metropolitan bishop, who keeps a tight rein over him, he cannot but express the opinions of his see, while his calling prevents him from giving certain forms of advice, as in the earlier example of the charivari.

The priests, therefore, now share with the *archontes* not the role of the *sophron* but the advice-giving function of those endowed with *sophrosyne*. One of the results of the advent of British rule in 1878 was to increase the number of lay teachers who helped harmonize, more cogently than earlier in the century, the ideals of Hellenic culture with those of eastern orthodoxy as constituents of national identity. The advent of the twentieth century thus witnesses the increasing social and professional differentiation of the highlanders, the growing ascendance of temporal values and the rise of the lay *sophron*.

The "sophron" and social change

The *sophron*'s role is also related to social mobility and the ranking system. The study of the transmission of plots located at the center of the village discloses that each "neighborhood," composed of a number of dwellings and of gardens, was originally the property of one man and that this neighborhood was continuously subdivided by inheritance, by gifts to sons, and the dowering of daughters. These central plots are prestigious in that they are considered as the heartland of the village and in that they serve to commemorate the name of those who founded the village in its present form. It is almost exclusively their present descendants in the male line who are said to come from an *archontogenia*, one of the lineages of notables.

The small number of "founders" in this very ancient village and their association with land do not appear to go further back than the eighteenth century. This may be due to the fact that the village, in the middle of the nineteenth century, was inhabited by only thirteen families as four deadly epidemics had, according to Alona tradition substantiated by contemporary documents,[17] visited the village in the seventeenth and early eighteenth centuries. During this period it was not unusual for Aloneftes to move to a new house built on a site which was situated at some distance from the one where misfortune had struck. Most of the priests of Alona who were also teachers – and all *sophrones* – whose names are remembered to this day are descended from these progenitors who, frequently, had also been the largest landowners and (most important in this community of farmers) its traders and money-lenders. They were, that is, members of an elite which owned, or which could raise through their network of wealthy patrons, the largest amounts of cash. Possession of this commodity provided them with a measure of freedom which was used both to influence the social and economic affairs of their village and to control its relations with its neighbors.

At the other end of the scale are the hired shepherds and the itinerant musicians. Both they and their descendants bear the stigma of these occupations. It would be disgraceful, especially for a woman, to marry one of them as they are said to be ignorant of the village rules of propriety. This has been translated into the accusation, made by their betters, that they are descended from strangers, *xenoi*, who came to live in the periphery of the village as handymen. Here, once again, the presumption of lengthy residence has been translated into high status and lack of status has been associated with alienness. This lowest social category also comprises the *mistarkoi*, low-paid unskilled wage-earners, often working in the asbestos mine or carrying out road repairs. It is only when these men are descendants of old established Alona families who pursue disreputable occupations in order to raise their children that no hereditary odium attaches to them. Thus, whereas the upper stratum of society constitutes the body of notables, the *archontoloin*, distinguished by the respect paid to those who are believed to have a long and distinguished association with the present village site, the lowest class, the *vilanarkon* (literally the villainous ones), are classified, and discriminated against, through the absence of that same criterion. Another collective name for them, *to phtocholoin*, the indigents, is revealing of the association in Alona understanding of alienness, uncouthness, and poverty.

The distinct social categories are those situated at the extremes of the scale. Using the criterion of descent and of respectability, the middle class sees itself as composed of those "honorable" families which can trace their descent from contemporaries, or near contemporaries, of those considered as "founders" of the village but whose ancestors, through lack of wealth or of association with prestigious office, did not make the same mark on village history as the forebears of *archontes*.

As the difference between *archontes* and "honorable" families is one of degree – and a tenuous one at that – the two classes intermarry, the lower one, ideally, providing daughters (rather than sons) to those above them, together with the largest dowry they can "raise" so as to mark the elevation of these brides to a higher social rank. The members of the middle class resident in Alona are either upwardly mobile or static, as they would consider it a disgrace to marry or to associate with those below them. In order to preserve the honor of their name, men in the middle class who cannot at least maintain their position in the village are forced to expatriate themselves. Their ideal is to return only when they have acquired sufficient wealth to be welcomed as grooms by an *archontike* family, this being, together with donations to their village, the ultimate consecration of their good name after years of labor abroad.

We have seen that the ideology of the social system rested on relative antiquity of descent, since prestigious office was either the apanage of the higher class or served, side by side with marriage, to elevate those nearest in origin to this class to a higher level. The ideal self-image of this closed society approximated to reality until the end of the nineteenth century. The breach of village isolation was followed by urbanization and emigration which opened new avenues to wealth and power while new elective offices[18] were created in the village. In time it became clear that men of low birth could convert their city-gained success into a prestigious village office so as to become the equals, or the superiors, or men of "honorable" families. Another possibility would have been to try to ignore the new social forces. This would have been fraught with even greater dangers for the newly rich, untrammeled by close bonds with the existing social system, might have created a new hierarchy of prestige based principally on wealth. This would have enabled those using it to set themselves above their society or, at least, beyond its social values.

Two remedies are traditionally used by Alona to counter this hazard. The first is to deny – or hinder – access to important offices to those who strive to convert talent, or newly acquired wealth or power, into status. This form of exclusion has been used against misfits: radical innovators, city rather than village orientated men, and husbands of "alien," in the sense of non-Alona, wives. The second serves to absorb rather than to reject. It consists in the use of a system of "communicating vessels." In this case those whose pursuits have met with exceptional success are selected for prestigious office so that, through their newly gained status, they are consecrated as social equals in relation both to each other, and to those who already occupy similar exalted positions. This arrangement facilitates communication between new achievers and traditional notables and integrates the successful into the village status hierarchy. It also serves to unite the village top echelons in defense of their shared privileges and, in cases where the same person is appointed to several offices, it helps to avoid conflicts between spheres of competence. This method of integration is applied mainly to "middle-class" Aloneftes, the members of the "honorable" families who, having made their fortune abroad or in the city, have returned to lead a traditional life in the village of their birth.

The concrete problems facing the notables of Alona after the end of the Second World War, problems which tested the Alona checks on social mobility and the *sophron*'s wisdom, were both social and economic. At the time traditional agricultural products were no longer marketable. Wine was being turned – at great loss – into alcohol, almond trees were uprooted while the hazelnut bushes, traditionally associated with this area since the days of the Lusignan kings,[19] were neglected. This was also the time when a number of Alona emigrants returned to their village, bringing back with them both capital and novel ideas concerning the feasibility of cultivating new crops. Where these had been successfully tested the emigrants endeavored to convince their fellow-villagers to follow their own example so that, eventually, a growers' marketing association could be formed. It might appear that the main interest of those involved in this venture was the expectation of its financial success. In reality the emigrants' stake was both economic and social. The sharing by them, within the proposed association, of prominent positions with notables would have enhanced the status of the emigrants of lower-class origin and, in this context, assimilated them to those descended from honorable families, thus opening the way to the marriage of their daughters to men of a previously inaccessible class. This, I believe, was the emigrants' most fervent wish and their ultimate objective.

The debate concerning issues and commensurate action was open and it continued over several years. This occasioned one of the widest rifts in public opinion that the village had known. The traditionalists feared all innovations. They pointed to the urban way of life of the "newcomers," to the "alien" wife of one of their number – a woman who, shamelessly, wished to open her own catering establishment in the village – and to the "forwardness" (*xetsiposia*, literally "lack of skin") of their daughters. To enter into an association with these people might change the face of the village, debase its values and lower its standing among its neighbors.

In cases of mild internal dissension the attitude of the priest, teacher, and most other office holders is relatively predictable. Whenever public opinion tends to be united, when it shows signs of following the lead of the *archontes* or the *sophron*, their assent (unless contrary to their calling) reinforces the decision and their own position within the community. When opinion is radically divided, each of the office holders follows the dictates of his estate and of his superiors. He may also, as in the present case, abstain from the discussion for fear of undermining his own authority.

The *archontes* are relatively freer agents, unless their interests are directly involved. Their power to influence opinion draws some of its strength from the heredity of their prestige. Although, in Cyprus, achievement counts for more than ascription, the *archontes* in the more conservative villages are listened to with respect as their words carry the weight of the good name of their ancestors. Members of respected families are also believed to be less affected by envy than lesser and more openly competitive men. In the present case the *archontes*, some of whom had successfully cultivated the new crops, were divided as to the expediency of forming an association with men of the lower class.

This was the problem faced by Hajjiyannis, the *sophron* of Alona, in the early 1950s. Hajjiyannis, like all *sophrones* before him, was descended from a prominent family. This served to allay the suspicion that the *sophron*'s advice might be inspired by a wish for self-advancement. At the same time his recognition as a *sophron* was contingent on the belief that he could not defend sectional interests, in this case those of his own class, without forfeiting the community's trust.

The advice of the *sophron* was to apply what I have described earlier as the second of the two Alona remedies to indiscriminate social mobility. It consisted, in this case, in the gradual absorption of successful returned emigrants through their appointment to respected offices, some of which would have been connected with the marketing association they had themselves advocated. This recommendation was both realistic and courageous. Realistic as its acceptance would have channeled new wealth and talent into the community. Courageous as the *sophron* was seen, at one and the same time, not only as siding against most members of his own class, but as the advocate of radical innovations; in this case of the initiative of members of the lower class to form a new association which, in due time, would have elevated them to a higher status.

This was the last important advice to be given by the *sophron* and the last stand of a coherent ruling class or, perhaps more aptly, of a class conscious that it was about to lose its coherence. Hajjiyannis died before his advice was accepted. The re-emigrants left the village once more and established themselves abroad. It was only then that the new crops gained ground and that the marketing association was consolidated.

It was only when one class of protagonists in this conflict between two systems of values – the traditional and the modern, the rural and the urban – had withdrawn from the fray that change was allowed to take its course. It was, that is, not so much the two "systems" but rather the interests of their representatives that were incompatible. Here the *sophron* failed to function as he was called to reconcile values which did not form part of a common scale.

Honesty, honor, wisdom, and sin

So much of Alona experience, and of its outlook on life, is expressed through anecdotes and words of wisdom that I shall let them preface my remarks.

Aloneftes, on being told that angels are sexless, feel no envy, and need no food, answer that although this is true of the Kingdom of Heaven, the first two characteristics pass human understanding. As for the third, it applies to the cicadas.[20] They sing but appear not to eat. Traditionally a long silence follows a thoughtful statement. This is necessary to help focus attention on the second delivery of – conclusive – wisdom. "Our old men have the wisdom of the cicadas. Having had their fill of everything they now sing 'hosannas on an empty stomach' [a reference to the meager diet of dependent elders]."

And again, "to live you need money and money is procured through sin." A man cannot function as a physiological and social entity without sinning. Sinning is a constituent of living. Only asocial or disembodied beings may remain sinless. They, and the elders who have withdrawn from social struggle and whose desires have wasted with their bodies.

Sophrosyne, in relation to honor, consists in steering a course that allows an actor to gain his ends without losing his reputation. To be accommodating, and thus suffer another man to prevail, is the sign of a *koroido*, a risible person, a dupe, an indelible mark not only of effeteness but of lack of judgment. A *koroido* is compared to a gelding. His meekness when wronged destines him to be a loser and a cuckold, as no woman can bear for long the disgrace of lying with an unmanly man, and no man can resist taking advantage of a fool. At the same time a man may be kind, even munificent, when his actions are motivated not by weakness or foolishness but by deliberate decisions resulting in means successfully matched by ends, as when alms are given to the poor or donations are made to the Church or the school, for these are not gratuitous actions but deeds which are expected to bring their own reward.

Are there, then, no other standards – apart from personal honor, success, and village reputation, prudence, and wisdom – by which an action may be measured? Honor and honesty share a common characteristic. They are less closely related to the "nature" of the action itself than to the moral distance between the actors, to the evaluation, that is, of the strength and character of the bonds linking them. Where there are none, questions of honor and honesty should not arise.

The following case, a *cause célèbre* in this area, may throw some light on the relativity of social values, and of the frontiers between shame and sin; that is of evaluations of conduct based on the nature of social relations as opposed to universal prescriptions edicted by divine law.

A young city woman returned to her home at the end of her studies abroad. She then joined, as an executive secretary in a nearby city, the firm of a wealthy businessman who became her lover. The man was of highland origin so that people of this area took an active interest in this affair. In due course the protagonists were cast into stereotyped roles. The businessman's success with his secretary, a well-born woman, was seen as the victory of village hardiness over city foppishness and laxity, while he became the avenger of village maidens who had been defiled by their young masters while working as servants in city homes. As for her father, basically a good man, he had been dishonored by his daughter who continued, unhindered, to heap disgrace on his name. He was, therefore, cast in the part of *koroido*, that of a risible dupe.

Sex, outside wedlock, results in honor gained and honor lost, man's gain being woman's loss. This is particularly true when social distance is markedly wide, as man, ever sensitive to the dictates of honor, adds a social dimension to his good fortune when he overcomes a woman of higher rank, while feminine sexual motivation, being mainly conditioned by physiology and sentiment, is less sensitive to the social evaluation of her lover. In terms of this masculine logic the sexual act is, for woman, an almost asocial fulfilment of her nature while, for man, the blend of physical and social achievement is measured by the qualities he has ravished: virginity, youth, beauty, and social position figuring among his trophies. The masculine imagery of sex is that of a hostile act: a battering-ram shattering the defenses of a castle, the impaling of an enemy, the planting of a flag on conquered territory. The subduing of modesty also forms part of man's strategy, so that each concession, seen as the infliction of a humiliation, marks a new victory.

The behavior of the lover, a natural braggart, was inspired by the belief that men want to seduce virgins while young women wish to seduce seducers. He therefore paraded his experiences freely, the stories becoming more scabrous with each telling.

The unfolding of these events was eagerly followed by the villagers. While the girl preserved her virginity, during the opening stages of this affair, the prevalent view was that, although a "true man" should be the guardian of his employee's honor, this widely traveled – and therefore knowledgable – girl could not but have led him on. Although their intimacy had defiled the girl, this reflected no shame on her lover. When the affair took another turn, the young woman was said to have been destroyed so that she was now, physically and morally, beyond repair. Nevertheless even the *sophron* opined that if this girl had really lent herself to the privacies described by her lover, she must have anticipated their natural conclusion. The lover, once again, was said to be free from blame. Thus intimacies with a teenage employee, followed by the loss of her virginity, were blamed entirely on her as long as the villagers' model of social distance and moral responsibility favored the lover (one of us) rather than his mistress (an "alien"). This is not true of sinfulness which is seen as a transgression of rules pertaining more closely to the human condition, and thus to universals, as opposed to local social regulations.

It is, I believe, the persistent denigration of the girl's reputation, the realization that she now had but little hope of marrying and settling down – indeed that she had paid dearly for her dishonor so that it would be a sin to inflict further damage – that provoked a reaction inspired by charity and pity. Now the "moral distance" factor operated no longer. The girl and her father ceased to be stereotypes. They became creatures of flesh and blood akin to members of one's own family – co-sharers of the human condition – on whom ill-fortune or God could inflict the same trials. "*Krima to koritsi!* a pity for the girl,[21] *amartia o geros!* [what] a sin [to inflict such a disgrace on] the old man! It is not for man to destroy what God has created." On attaining the frontiers of sin, we have reached the limits of the *sophron*'s sphere of competence.

The social conflicts that I have described point to the multiplicity of models each composed of the same reference (e.g., nation, state, community, family, etc.) disposed in a different ranking order.[22] Thus there are two categories of model. The first is considered as impersonal, and its ranking order of values remains constant. On the temporal plane this is the counterpart of the divine order.[23] The function of this "static" category is to provide, on its own plane, an ultimate scale by which conduct may be evaluated. The second category contains a profusion of relative models, each with its own ranking order, the ranking being adapted to various social contingencies such as age, sex, social position, etc. at various stages of life. The relative models allow for pragmatics and for the justification of conduct at variance with the transcendent "static" model. It also makes possible the malicious faulting of a countryman's motivation.

An actual example may illustrate this point. At the outbreak of the war a man volunteered to serve in the army and was killed in action. His community erected a monument to commemorate an act which upheld an absolute ideal, that of putting country before self, in this case even before family as the man was married and had children. Envy immediately raised its head. The soldier was said to have sacrificed his wife and his children to vainglory, a choice that no sensible man would make at his stage in life. His own unmarried kinsmen, shamed by his action, spread rumors that the dead man's wife was such a termagant that joining the forces was preferable to living with her, so that the choice, in this case, was an egoistical one. Each of the motives attributed to the volunteer was part of an evaluation corresponding to its own scale of values.

It is true that each society uses its own methods for evaluating conduct in relation not only to its ideal, but also, to its conception of the reasonable man. What is particular to the social pattern characteristic of these Greek Cypriot villages is that the individual and the family – even the national culture – often take precedence over the state, and that collective opinion beyond a very narrow range of kinsmen and of members of one's community is relatively inoperative. Thus relations beyond these narrow social confines may be compared to international relations when confrontation of interests is, in the final analysis, mediated by a vague *jus gentium* and natural law. As the *sophron*'s task is to mediate between ideals and reality, it is the flexibility of the ranking order of values and the casuistry explaining a choice that bestow exceptional importance on the *sophron*'s function within the community.

At the same time it would be a travesty of reality to consider the life of the Pitsilia highlander as limited by considerations of gain and aggrandizement. His egoism rests at times and his mind is illuminated by a dream to which both his national identity and his transcendent model of values are attuned.

The Greek Cypriot lives in so many dimensions of time that otherworldliness, here in the sense of a world that has been and that will come again, is a common experience. He, the poor Greek Cypriot highlander, sees himself as the heir to classical Greece – and of Achilles and Nestor, of Pericles and Alexander, of the imperial glory of the Orthodox court of Byzantium whose ceremonial is re-enacted in his mountain church. Christ and the last Palaeologue were martyred to make him a man by instilling into him a faith that makes life worth living. Twice did he fall from grace. When Adam was turned out of Paradise and when the Polis fell in 1453. The conquerors that have been overrunning his island have done no more than enact this desecration. But just as surely as Christ rose from the dead the Emperor, now turned to marble, will feel his pulse quicken again and *e Polis* and Cyprus will come once more unto their own until the time when the trumpet of the seventh angel sounds the end of all earthly kingdoms.

For the time being the Pitsilia mountaineer will continue his daily struggle using *sophrosyne* to keep body and honor together while he preserves his soul, with the help of his Church, from being weighed down with too many *krimata*. As long as he keeps alive and undivided within his breast the Hellenic, the Byzantine, and the Orthodox faith, whatever his *krimata* but not his dishonor, he cannot fall from grace. This is his *riziko*, his destiny. Whoever speaks to him while appealing to these values is listened to with respect. This is the task of the *sophron* who, working hand in hand with the priest and the teacher, defends the good name of the village and of its inhabitants and does not allow evil to triumph.

Notes

1 Field-work was carried out in 1954 and then at various periods between that year and 1983 with the generous help of the Nuffield, the Ford, and the Wenner-Gren Foundations. I would like to acknowledge my indebtedness to M. Clemens Heller and the Maison des Sciences de l'Homme for inviting me to Paris to participate in

some of Professor Julian Pitt-Rivers' seminars during which parts of this paper were discussed.

2 Over twenty villages situated in the eastern range of Mt. Troodos (and not the western range as is, mistakenly, mentioned in *Honor and Shame*) are collectively known as the region of Pitsilia.

3 There is a confusion between *stroma* and *krevati* whose respective meanings in mainland Greek speech are "mattress" and "bed." This may be due to the fact that, in these highland villages, there were no proper beds made and used exclusively for sleeping but that mattresses were laid, at night, atop low tables which in daytime had other household uses such as frames for the shelling of hazelnuts.

When the *karkola*, in the form of an iron bedstead, was introduced in Alona it was evident that its sole function was that of a bed. It was, thus, clearly differentiated from other pieces of furniture which, although used as beds, could also serve other domestic needs. The modern bed thus acquired strong sexual connotations and became associated with moral laxity.

4 *Aien aristevein kai ypeirochon emmenai allon.*

5 See p. 116.

6 The coffee shop, *kafeneion*, is the centre of male village life. A forthcoming publication by Papataxiarchis describes its multiform social functions.

7 The *archontes*, the leaders, the masters, are village notables who owe their rank mostly to their descent (mainly in the male line) from respected ancestors, an *archontogeneia*.

8 It is increasingly difficult to differentiate between residents and non-residents as many men, considered as residents, are absent during substantial periods of the year. Here only those who reside in Alona (but, possibly, work in the city) are considered as residents.

9 The same residential criteria as those of note 8 apply in this case.

10 *Linobambakos*, literally "linen-cotton," has the meaning of "neither fish nor flesh." The word was used in relation to a small number of Cypriots usually living in remote areas who, under Turkish occupation, behaved at times as Christians, while under pressure they claimed to be Muslims.

11 It is considered that the performance itself reflects such shame on the actors, especially on the one playing the female role, that they had to be "imported" from the lowlands.

12 A village endowed with a church and a resident priest.

13 The completion of this survey and the analysis of the data was interrupted by the Turkish invasion in 1974.

14 In order that their secrets might best be preserved.

15 Makhairas is one of the oldest and most prestigious monasteries of Cyprus.

16 As all secondary school boys, that is over 94 per cent of village boys over the age of 12 or 13 attend school in Nicosia, the capital city, where most of them had to reside when communications were uneven, the male urban sexual model has been adopted. 1911 marks the year when the first Aloneftes went to school in Nicosia to be followed, in 1946, by the first Alona girl. Nowadays practically the same proportion of girls attend school in Nicosia but for them virginity remains one of the essential prerequisites of marriage.

17 Mainly the local church records.

18 The majority of these are connected with the powerful local banking and commercial cooperatives.

19 A French noble family from Poitou which acquired the island in 1192 (during the third crusade). The Lusignans ruled Cyprus until 1489. During this period Cyprus was the principal "Latin" center in the eastern Aegean.

20 The French *cigale*, in English cicada, is often mistaken by English authors for a grasshopper. Somerset Maugham has written about "the ant and the grasshopper" and we are told by the *Oxford Companion to English Literature*, under La Fontaine, that his fable "La Cigale et la fourmi" (hated by children, all over the world, who have to memorize it at an early age as part of their French lessons) is about the "grasshopper and the ant," which makes nonsense of the whole story. In any case La Fontaine got his facts wrong as the cicada spends most of its existence (of over ten years) underground and only emerges from its underground habitat to make the shrill sounds described by La Fontaine as a song when the time has come for it to reproduce and die. Not only is it unnecessary for the cicada which draws its substance, when underground, from the roots of its host tree, to store food for a winter which it will not survive, but the cicada, during its life above ground, needs no food.

21 The main meanings of *krima* are those of sin, trespass, lack of respect, of an unjust or criminal action, and of pity. Here it is used in the sense of sin, and of pity for someone whom fate has treated harshly.

22 I have referred to the multiplicity of models, of the struggle between macrosocieties and microsocieties, both in the Introduction and in the article concerning this area in J. G. Peristiany (ed.) *Honor and Shame: The Values of Mediterranean Society*, Weidenfeld and Nicolson, London, 1955, and also on pp. 24–5 of my Introduction to J. G. Peristiany (ed.) *Mediterranean Family Structures*, Cambridge University Press, Cambridge, 1976, concerning Antigone's dilemma, the difficulty of choosing between different scales of values, which forms the subject both of many Greek tragedies and, on a less exalted plane, of the "difficulty of being" in contemporary Greek society.

23 If, for Emile Durkheim, the divine order is a refraction of the social order, for the faithful, and this includes the Pitsilloi, the social order is a debased version of the divine order. It is, therefore, possible to fault any decision based on Caesar's model by reference to that of God.

7

The Greek hero[1]

JOHN CAMPBELL

For Yannis Koliopoulos

Brigandage was endemic in the mountains of continental Greece throughout the period of Ottoman administration. The Greeks who engaged in this activity were generally young bachelors belonging to the mountain villages and transhumant shepherd communities in the area where they operated. They shared with the inhabitants of the mountains, who were also their principal victims, the same social values predicated on notions of honor, pride, and shame; and the same social institutions which protected the interests and reputation of the family. The role of the young unmarried man was that of the guardian and, if the need arose, the avenger of family honor. Indeed, as we shall see, involvement in an honor killing was one situation which might convert a young mountain villager or shepherd into a brigand. This category of young men, whether or not they became brigands, were expected to possess and display heroic qualities of courage and physical strength. It is these qualities and their attribution to brigands (or klephts as they were known), in the eighteenth century, that I wish to discuss.

I shall also examine, very briefly, the position of another class of social heroes, the Homeric warriors of the *Iliad*. There is no intention to claim any direct relationship between the two societies. The circumstantial differences are obvious. There are, however, common features that direct attention to patterns of behavior which appear to be more than contingent aspects of this form of heroic society.

Although no doubt some features of Minoan and Mycenean culture persist in the *Iliad* and the *Odyssey*, the social world which Homer there lays before us is that of the Dark Age in which small communities faced a hostile world where men steal and plunder, as much for prestige as for subsistence, and rely only on the trust and support of kinsmen and dependants. Although it is a culture with a shared system of values, and complementary conventions of hostility and hospitality, its constituent social groups, households, communities, petty kingdoms, were worlds with sharply defined moral and physical frontiers (Redfield 1975: 99).[2]

The leading warriors of these kingdoms are described by the Greek word *heroes*, heroes. These were the men with armor and with chariots, the *promachoi*, the protagonists in the battle. In Homer heroism is a social role. Its task is the defense of the community's limits; these may be physical, as are the walls of Troy, or the wall round the Greek camp, but they are also moral. The hero's actions reflect on his community's reputation, but they also refer to his personal honor. The ambiguity and disorder which are the source of the dramatic action in the *Iliad* arise precisely in the relation between personal honor and the honor of the community. The very qualities which the hero must rely on to carry him through the trials which face him are those which in other contexts may threaten the stability of the group. The character and experience of Achilles reveal the dilemma. Agamemnon, as leader of the army, demands for himself the captive girl Briseis, taken by Achilles. Having made this provocative claim it is impossible for him to withdraw it. In Homeric society authority and status are confirmed only by their exercise. Achilles understands this but his awareness of what is due to himself must be his single concern. Honor is a claim, and one justifies it only as long as it is not challenged. It is permanently at risk, and there is no possibility of compromise. The rage, *cholos*, which now consumed Achilles would have driven him to kill Agamemnon had not Athena intervened. The remaining alternative for Achilles was to withdraw to his tents, there to make himself entirely independent of those who had wronged him. This withdrawal removed from the battle the warrior on whom the Greeks principally depended for victory; but it also assuaged the offense to Achilles' honor, by establishing his independence. He owed nothing to anybody. The embassy which comes to him from Agamemnon "with glorious gifts and soothing words," therefore, had no hope of success. By offering the gifts of seven towns, and his own daughter, Agamemnon is making a display of his resources and his status. In effect he is offering Achilles the dependent position of a son-in-law. Achilles says that Agamemnon has treated him like "a wanderer without honor," precisely the position of the man who marries into his father-in-law's home in many parts of the Greek countryside to-day. It is only the death of Patroclus that draws Achilles into the battle once more, under the necessity of avenging his companion, an obligation of personal honor, and not in any sense as a late recognition of his duty to the Greeks.

Achilles is the extreme example of the hero. Like every Homeric figure who wishes to claim that status, Achilles boasts. In the fiercely competitive world he inhabits, not to do so would betray a lack of confidence in his powers, the proof of which lies always in the battle yet to be fought (Redfield 1975: 154). He is stronger, braver, more eloquent, more brilliant than the others; an exceptional being, prepotent, a force of nature hardly to be contained within the constraints of his role. Particularly after the death of Patroclus, he kills his opponents with a cold and furious facility. The exceptional qualities of this hero transcend merely human powers. Achilles is the son of a goddess, Thetis, and a human father, Peleus, and his mortality is therefore mixed with divine powers. It emphasizes his ambiguous role, part man, part god, with the power to ensure victory to the Greeks, yet, because of that power, fated through his sense of honor to spread disturbance and disorder.

Hector, Achilles' principal opponent, is also in the Homeric sense of the term a hero. He is, however, a hero on a more human plane, although like other heroes he receives divine support at particular moments in the struggle from both Zeus and Apollo. Born to a princely status he is a model of courageous and honorable virtue, qualities he has learned by responding to *aidos*, a sense of shame (Redfield 1975: 115; Adkins 1960: 48ff.), which makes him sensitive to the judgments of others. If Achilles pushes his natural forces to the limit in the assertion of his honor, Hector is the hero who explores the limits of his courage in defending his city and his family. Hector leads the Trojan army; and he has a family. Achilles is not burdened with these potentially conflicting roles. Hector is caught up in a complex set of relationships with Priam and Hecuba, with Andromache his wife and their small son, with Helen, with his brothers, with other leading Trojan warriors and the allies who have come to assist the beleaguered city.

In deciding a course of action Hector attempts to tread his way with care, balancing the claims of different obligations and responsibilities. As a warrior he must react to a challenge without pause for consideration; as a mature leader of his army he ought to show prudence and even consider ordering a tactical withdrawal if the situation requires it. He is therefore a man at times not without anxiety and even fear. In the last analysis we see that he is vulnerable to the call of the warrior's honor. After the day of victories which Zeus deceptively grants him, he finds it impossible to accept the cautious advice of Polydamas that he should withdraw the Trojan forces within the walls of the city, vastly outnumbered as they were by their Greek attackers. Later, again, when he stays alone, outside the walls, he ignores the pleas of Priam and Hecuba that he should withdraw.

These decisions lead inevitably to his death. This, too, we know from the legend is soon to be the fate of Achilles, as indeed it is the presumptive end of every hero. The hero in youthful invincibility appears to be godlike but in fact he is mortal (Redfield 1975: 101). In guarding the moral and physical boundaries of his community his existence is permanently at risk. For this, however, there is a reward which carries weight in a society whose principal value is honor, for although he cannot survive physically, his fame will be preserved, sung, and recited forever.

There is one fear, that the enemy will possess his body, despoil and desecrate it, and prevent his companions from giving him proper burial. Why does this matter? It matters because honor, like ritual, is principally a problem of external form and effect, not of intentions. Physical appearance, social decorum, the integrity of limits are constituents of it. Thersites, who speaks in a disorderly fashion throughout the poem, is appropriately ugly and physically deformed. To mutilate a dead body is to make it *aeikes*, unseemly, and therefore dishonored (Segal 1971: 15). The values of Homer's heroes, like any relatively consistent system of social norms, support an ordered pattern of relations. In the earlier part of the struggle for Troy the enemy dead, stripped of their arms, were returned to their own people for burial. The loss of armor, which is an extension of the warrior's personality and strikes terror into his opponents, is serious and dishonoring; hence the importance of closing ranks to defend the body of the fallen hero. But these aspects of death and defeat are tolerable if the hero is given a fitting burial. For this is a reaffirmation of his reputation by his equals; in the rituals of the burial, in the marking of his tomb and, for the foremost warriors, in the celebration of their name at funeral games which link it with feats of excellence. Later, however, the ordered exchange of dead warriors was threatened, first when Hector, incensed by Patroclus killing his brother Cebriones, tells the dying Patroclus that he will be treated as carrion, that the vultures will eat him. This intention provokes the furious battle over the dead man's body. Iris also tells Achilles that Hector intends to cut off Patroclus' head and fix it to a stake. The possibility that a defeated hero may be decapitated, or fed to dogs and birds, is an extension of unsanctioned violence which takes matters beyond accepted conventions. For the Homeric warrior it is a terrifying fate. The defilement of the corpse and the denial of burial represent a mark of dishonor which implies the extinction of the victim's name in the memory of this world and, in the world of the dead, a sentence of exile for the soul of the dead man which must wander eternally outside the gates of Hades.

The exclusive commitments which honor demands, and the extreme difficulty of retreating from a claim advanced, or a threat made in the heat of battle or dispute, constantly threaten Homeric society with moral disaster, the cause of which lies in the very heroic qualities on which it is based. Those who in a sense have the most to fear from social disorder are the women. In the Homeric household, women are permanently *in loco filiae*. If taken in war they, with their children, become the property of the captor. They are wholly dependent on the strength and protection of their warrior husbands. Their task is obedience and fidelity, the symbol of their integrity. If they fail in these respects they undermine the order of domestic life which rests on male authority. Hector's Andromache is the model of the faithful wife who complements the hero. She is contrasted with Helen, a source of danger and indeed cosmic disorder. The fact that Helen herself is not intrinsically evil, for she has not actively planned to bring about the situation in which she finds herself, only accentuates the threat of undisciplined sexuality which she represents.

The outward nature of the woman's role is passive, enduring, not heroic. Yet she is the mother of heroes and her reputation bears directly on the honor of her son, for although the son succeeds to the public personality of his father, it is from his mother that the hero draws those powers which mark him out individually. In this identity with her son, and in her dependence on her husband, she may in a crisis become a fierce censor of their actions. Cowardice or apathy threatens honor and, indeed, the survival of the community, whose destruction brings an end to her world. In Book IX the wife of Meleager has to encourage that warrior to take the field by a description of all the consequences of defeat for the city (Redfield 1975: 120).

In the Homeric household, men and women are close; their roles are different but complementary. In war, however, they are separated. Within the walls the children, old men, and especially the women, fearfully await the outcome. Troy stands for the social order of civilized life threatened by the consequences of war. When Hector has already been killed, his wife Andromache, in ignorance, is still attending to her household duties. She is weaving "in the innermost part of the lofty house" (XXII.440). She orders a hot bath to be made ready for Hector when he returns from the battle. Eventually she learns of Hector's death from the distant anguished cries of his mother. From the walls she sees the despoiled body of her husband dragged behind Achilles' chariot. In her agony she imagines the body given over to dogs and worms. She says she will burn the "light and graceful" garments in which Hector would have been laid had he been given the honorable funeral to which he is entitled (Segal 1971: 46–7). By these details Homer emphasizes both the intimate relationship between the warrior and his wife, and the contrast between the inner world of domestic civility and the realities outside the walls of Troy of an increasingly brutalized conflict. The women have no illusions about the consequences of defeat. Before Hector takes the field for the last time (VI.407–11) Andromache remonstrates with him:

> You strange creature, your strength will fail, you've
> No pity for your son, for me, ill-fated, soon your widow,
> As I shall be. Soon they will kill you, these Achaeans
> Massing against you. And then I'll find it better
> Once I have lost you, to sink within the earth. (Redfield 1975: 123)

And later (XXII.491–8), when Hector is dead, she gives this account of what it will mean to her son to become an orphan:

> Ever he hangs his head, the tears run down;
> In lack the child goes to his father's friends,
> Plucking the coat of one, and another's shirt.

> In pity one hands him a drop in a cup
> To moisten his lips, but his mouth it doesn't moisten.
> At the flourishing feast, too, they insult him
> With a blow of the hand, and the harsh sting of an insult:
> "Go along, Your father does not feast among us." (Ibid.: 125)

The demands on heroes are insistent and continuous. Although their social position entitles them to privileges of property and status, they only validate these claims by displaying the virtues expected of a warrior. Before the battle they must boast of their prowess; in the contest itself they must prove that they have the strength to face the enemy. Their preoccupation with combat makes it unlikely that they fear the physical shock of violent death; but their terror of mutilation after death, defilement, and denial of honorable burial is acute.They are also anxious about the fate of their families and dependants in case of defeat. Even the foremost heroes know fear. When Hector faces Ajax, his limbs tremble. The hero appears godlike to his dependants, but he knows better than others his reliance on the support of the Gods. It is true that some heroes, Achilles, Ajax, Sarpedon, for instance, have one parent who is a god or goddess, and that they draw strength from this divine descent. As a class, heroes stand between ordinary men and the Gods of Olympus. In the press of battle, however, in combat with equals, this may not be enough. In the *Iliad* the Gods regularly intervene in the outcome of personal contests, either to destroy the opponent of a favored warrior, or to save him from death. As we have seen, the hero must boast and show himself to be supremely confident. Yet, at every step his insecurity is evident. The word, *euchesthai* (Redfield 1975: 129), to pray, is also used in the sense of to boast, or to boast vainly. It is unwise to boast without at the same time making supplication to the Gods. Throughout the poem heroes offer prayers, vows, and sacrifices (Griffin 1980: 148), to win the favor of Zeus, or some other god. The search for glory, and the fear of extinction, are sentiments and activities which dominate the existence of every hero, even Achilles. Together they determine the need for Grace.

Keeping in mind these features of the Homeric world which define the hero's position as a social role, the task of a class of leading warriors, we turn to the world of mountain villages and shepherd encampments in the Pindus mountains in continental Greece during the eighteenth century.

At the beginning of that century the Ottoman administration of what is today continental Greece north of the Saronic Gulf employed bodies of locally raised militia, mainly recruited from the Christian population and known as armatoles, to keep order, arrest criminals, and suppress brigandage. They also guarded bridges and passes on strategic routes in this extremely mountainous area. All traffic, one must remember, State officials, collectors of revenue in kind and in money, and the goods of internal trade, were carried on mules and horses along vulnerable tracks which passed through deserted and broken mountain country. A typical incident is celebrated in a song:

> The muleteer descends from the mountain peaks
> twelve mules with twelve judges he leads down
> and there straight in the road the klephts await him. (Politis 1973: ιβ')

The song has its tragic end. The leading brigand stabs the muleteer only to discover it is his own brother. But this is the characteristic breach of order which armatoles were recruited to police. Bodies of local militia had existed in the Byzantine Empire, and they re-appeared as armatoles early in the Ottoman period. Their opponents were small groups of brigands known in this period of their history as klephts, a word which in one sense simply means thief, but carries more complex associations in patriotic claims to heroism and sometimes counter-charges of brutality.

Brigandage in the Pindus had its roots in the social structure of mountain villages and the encampments of transhumant shepherd groups which rented their pastures (Vasdravellis 1970: 27–9). In both, the family, whether in its simple or extended form, was the social group which commanded the almost exclusive loyalty and attention of its members. It was not simply a domestic association of individuals with affections based on blood relationship, but a corporate enterprise working together to provide for the subsistence of its members by the exploitation of its fields, flocks or other resources. It was also a religious community (Campbell 1964: 37) protected by its ikons and other sacred objects. The house itself, like the Homeric *oikos*, was a sanctuary for both its members and its guests. To this institution a man owed an exclusive duty which seriously limited the possibility of friendship to outsiders or disinterested membership in other social groups. There was a fundamental division between "our own people" (*dikoi mas*) and "strangers" (*xenoi*), marked by attitudes of intense secrecy and distrust which protected family interests by an intricate screen of justifications and lies. In these communities families continuously competed for social preeminence, for a prestige in which moral reputation and material resources were closely linked. The poor were not blessed in any sense, and reputation without the material basis to support it was scarcely possible. Each family carefully cultivated its self-interest (*symferon*) whose boundaries were marked by the perimeter of its honor. The defense of this honor depended on the complementary sex-linked qualities of *andrismos* (manliness) and *drope* (in it sense of the sexual shame of women). These were the qualities which enabled men and women to fulfil the expectations of their roles and duties in the service of the family. For all the differences of time, scale, context, honor here was close to its Homeric analogue, expressing a sense of integrity and independence from constraint, of not being humiliated before others, or touched by particular kinds of failure or loss. Under a regime of such values peasants and shepherds looked on their local world as a closed economy of fixed resources. What was owned by or allocated to one family, for instance in animals or grazing rights, was denied to another.

In these mountainous areas both villagers and shepherds depended on flocks of sheep and goats, and their hostilities were mainly expressed in the theft of animals (Vasdravellis 1970: 27–9; Campbell 1964: 206–9), not as a rule from immediate neighbors (for such exploits would be imprudent and morally ambiguous) but from flocks some miles away where distance lent some defense; but still within a district where prowess, cunning, and daring could be claimed and known, and would therefore act to delay, if not to deter, retaliation. Not every act of theft was identified or revenged, but the incidence of theft and counter-theft was a nexus of reciprocal hostilities which bound families into balanced patterns of alliance and opposition. Marriages were contracted with some thought about the support of affinal relations, the *sympetheroi*, who could be relied on to distribute and hide stolen animals in their flocks, or to identify in flocks of their immediate neighbors other animals which had been stolen from their kinsmen. Stealing animals is a dangerous occupation. Not only are there sometimes affrays with sticks and knives and dogs in the dark, but the victim of a theft is often driven by his indignation to voice his suspicions. Words, as we have noted in the Homeric world, are difficult to withdraw in questions of honor. Insult leads to homicide. Other matters unconnected with animal theft, also, might result in killings, especially anything which reflected on the chastity of the women of the family: the slander of a sister's virtue, a broken betrothal, or worse. The blood of a murdered brother or father is another category of wrong which called unceasingly for vengeance, whatever justification the offender might have had. But the family, even the extended family, was a weak and vulnerable group. Despite the support of kinsmen its membership was not adequate for the prosecution of extended feuds. It lacked the strength in numbers that might, for instance, be found in a Bedouin lineage group of even modest depth. Above all, it could not afford to lose its head, and leader, without whose presence and protection the position of the women and children was implicitly dishonorable. For this reason the task of vengeance killing, of rebutting verbal insult, of cutting out and stealing other people's animals, was the role of the unmarried son who, unburdened as yet with the responsibilities of a head of family, was able to concentrate his thoughts on the vindication of personal and family honor. These unmarried men at the height of their physical powers were known as *pallikaria* (Campbell 1964: 278–81). This term expressed the ideal of the hero warrior with physical strength and assertive courage in the very flower of his youth yet whose life at every moment was at risk in the defense of his honor. *Pallikari* was often coupled with another term, *leventis*, which described a youth who was handsome, manly, narrow hipped, quick on his feet. The aggregate of these qualities was the ideal of youth. It involved, too, a certain conception of innocence, perhaps of immaturity, of a man unmarried, not involved in sexual relations, in responsibilities and leadership, and in the lies and contrivances which, it was believed, necessarily accompanied them. By contrast the married head of family required prudence, foresight and some cunning. The parallel with the different positions and experiences of Achilles and Hector is sufficiently clear.

Yet the young man involved in an affair of honor who had killed, or expected to be killed because of acts committed or insults given, had no choice but to leave his family and his community. His own death, if he remained, would create a debt of honor, payment of which could only lead to a further killing and, in the absence of another unmarried son, this would destroy the family, or at best leave it weak, vulnerable, and potentially dishonored.

Men forced to leave their families and villages took to the mountains. This was a permanent source of recruitment for the klephtic profession. There were, of course, personal disasters created by the effects of wider political policies which forced other men to the same course of action (Vasdravellis 1970: 27). In the earlier period of Ottoman rule the periodical recruitment of Christian children (ibid.) to be trained for the Janissary Corps (a forced enlistment which removed them permanently from their families and their religion) was an assault on the sanctuary of family which many could not tolerate. Forced conversion of adults to Islam was another threat. Less general, but equally arbitrary, were the distraint of goods and land for debts or unpaid taxes, and arson and looting perpetrated by the klephts themselves. Any of these circumstances could drive men into the hills.

Klephts organized themselves in bands[3] generally numbering between twenty and thirty men, about the number which a leader without the support of hierarchical discipline could reasonably dominate by the force of his personality. When we read of bodies of klephts with as many as 400 men this is invariably a temporary confederation of smaller bands which did not surrender their identities. The leader of a band was known as the *capetanios* (captain) or *protokleftis* (the first klepht), his second-in-command was the *protopallikari*, the leading *pallikari*, and the remainder of his followers were simply *pallikaria*. The use of the word *pallikari* here points to a continuity between the young man's role, first as family hero and then as klephtic hero. If the captain were illiterate, as was generally the case, he might also have a secretary (*grammatikos*). There were also in many bands *psychoyoi*, so-called foster sons, adolescent boys serving an apprenticeship as servant and protégé of an adult klepht. The captain, his lieutenant, and the remaining klephts divided the proceeds of their operations in proportions which reflected their different statuses, but it seems that the foster sons were provided only with shoe leather and subsistence. The band in some respects resembled, and was modeled on, an extended family without the presence of women, the same kinship basis as that of the typical transhumant shepherd group, the *tselingato*, from which many klephts were drawn. In such bands there was generally a nucleus of two or three brothers, cousins, or affinal kinsmen, who provided the leader with unquestioning support. Between these men and the other unrelated klephts a network of ritual relations was created through blood brotherhood. In a society where unrelated men were mutually hostile and untrusting, the *vlamides* or *bratimoi* owed each other the trust and straight dealing on which their survival depended – in the same measure as brothers. Brothers shared the same blood which literally represented their moral equivalence. This was reflected in the ritual of blood brotherhood in which the blood of two men was mingled while a monk read the Gospel over them.

A band of klephts generally attempted to operate in an area known to the captain and his closest associates. Kinsmen in villages and shepherd encampments provided intelligence and food (meat, flour, and salt) without which they could not survive or plan rational operations. Nevertheless these men had been forced out of the village community, beyond its boundary guarded by the small chapels and ikon shrines which separated a space where men lived in civil society protected by saints and blessed by God. Beyond was an area of danger, a field of battle as it were, inhabited by snakes, bears, wild boars, and less material demons.[4] In this wilderness, the klephts created what order they could, establishing a number of camps, *limeria*, in inaccessible positions, between which they circulated to avoid surprise. As far as they were able, under the austere life they led, they paid great attention to their clothing and their arms on which the leaders especially were prepared to spend extravagantly. In common with armatoles and Albanian soldiers, they wore the white *fustanella* kilt, elaborately decorated waistcoats with silver buttons, and silver knee-plates when they could be afforded. Personal cleanliness was not a concern and it was a matter of professional pride that their kilts had become grey with encrusted dirt and grease. They armed themselves with long guns, curved swords with a cutting edge, and daggers. These could all be very elaborately decorated.

This concern with the display of pride, of appearing above others, was a form of visual boast, which also had the effect of discouraging their opponents. Released from the constraints of life in a community, they could now live by the more assertive features of personal honor. Whenever the strength of the opposition permitted, they liked to appear provocatively before villages with flags and banners flowing, and in some instances with trumpets sounding.

Forced to leave society because of the injuries they had suffered, or the disorder they would provoke, the klephts returned to prey on it. This, naturally, was the necessary condition of their survival in the mountains. The klephtic leader needed money to equip and decorate himself, provision his followers, and offer them some hope of a share in the surplus. It is clear from the documents that in a great majority of cases the victims were Greeks. When a band was firmly established, it threatened to burn a village if money and provisions were not delivered, or it might seize hostages for ransom. A monastery document of 1764 (Politis 1973: ιγ́ n. 5) records that three hundred brigands led jointly by a Greek Gioldaris and a Muslim Albanian descended on Karpenisi, "ill-treated the unhappy inhabitants, seizing all their movable property and leading away many hostages of whom some died in captivity and others after many months of misery and persecution were released for a large ransom."

An earlier document (1685) (Vasdravellis 1954: 88) tells us of a klepht, Nikos Spyros, who broke into a man's house, bound and beat him, and burnt the breast of his wife to make them show him where their money was hidden. Burning captives with boiling oil or butter was a commonly employed torture. Indeed the cruelty of klephts is well attested. Robert Curzon coming down the track from Metsovo to Jannina met a muleteer with a small packing-case tied to his saddle. Out of the case peered a white face, racked in pain. Curzon supposed he must be a dwarf until the muleteer asked him if he knew of a good physician in Jannina, since his passenger had recently had both his legs cut off by klephts on the way from Salonika. There is sufficient evidence to show that, if a delivery of a ransom was delayed or refused, threatened mutilations of hostages were promptly effected. Yet there is also evidence that klephts treated their captives well, shared their food, found them what comforts they could, treated them indeed with a friendly camaraderie, and respected the chastity of the women captives. The brigand camp, the *limeri*, was a house of sorts and was bound by the same laws of hospitality up to the point where the bad faith of the captive in not revealing where his valuables were hidden, or of his family in not delivering a ransom, cancelled that obligation. Then, the execution or mutilation of the captive was inevitable and indeed logical. A ransom evaded was honor denied, and that insult could be expunged in no other way. If a band was to operate and survive it was essential that it was feared and respected. The world at large had to be completely convinced that if the ransom was not forthcoming the captive would die. Generally, he was beheaded. Equally, once a ransom was delivered, the only concern of the band was to restore normal relations with society which the act of capture had suspended. In a ritual of reconciliation the captive was shaved, a lamb was roasted, and gifts were exchanged with the emissaries who had brought the ransom.

Klephtic guerilla tactics (Skiotis 1975: 315) were adapted to the two kinds of action they were most likely to face: the setting of an ambush, and a delaying action against numerically larger forces set to capture or destroy them. In the second and more serious instance they selected positions commanding passes or defiles, which their opponents could not avoid. Concealed from view behind rocks or shallow entrenchments, *tambouria*, they were sited to command a section of the track where they could concentrate fire and disorganize the enemy in the ensuing melée of dead and wounded men and horses. If the effort was successful, the sudden charge, *yourousi*, with swords might follow. They were extremely adept at moving from one position to another, using the cover of brushwood and rocks to maximum effect. Their agility and speed in climbing over broken and mountainous country was remarked by many foreign observers, among them Leake, Hughes, and Urquart. They could disengage from one action, only to pose identical problems for their opponents a few hundred yards further on.

Despite the absurd disparities in numbers engaged, and casualties inflicted on either side, which klephtic accounts of their exploits claimed, there is no doubt that as long as they did not venture into the open, they could delay vastly superior forces in the tangle of steep, narrow-sided valleys which characterise the Pindus mountains especially on the western flank. At first sight this mode of warfare may seem less than heroic; yet it is individual irregular warfare in which the warrior moved and decided his own path, balancing his courage, his prudence, and technical prowess with his gun. The safety and survival of the captain, whose name and personality stood for the continuity of the band, were a first concern and therefore his position was generally carefully concealed. Yet he had to inspire and exhort his men by his vocal and personal encouragement. As in the case of a head of family in the civil society of the village his loss threatened the dissolution of the group. And it was the task of the young *pallikari* hero, not the leader, to die if that sacrifice was necessary.

I must now return to say something briefly about the armatoles. From the earliest years of the Ottoman period (Vakalopoulos 1964: vol. I, 216), about 1430 in the case of the Agrapha district, they were recruited from the ranks of the more dangerous and effective klephts to contain banditry and keep the principal routes open to travelers. Normally, they were appointed by the Ottoman Cadi at an agreed salary, which was collectively paid by the district they protected. In the seventeenth century we have a number of documents which illustrate the handing over of captured bandits by armatoles to the Turkish authorities for judgment, and others in which the local inhabitants take the initiative in asking for their appointment (Vasdravellis 1970: 40–50). Eventually the whole mountainous extent between Macedonia, Epirus, and the Saronic Gulf was divided in a number of armatoliks[5] each of which was subdivided into smaller commands. Armatoles were sometimes granted important privileges, exemption from taxes, and an undertaking that regular Turkish forces would not enter their district. The possibilities of exploitation hardly need to be stressed in an administrative system where it was accepted that public office was a means to personal fortune. An imperial Firman of 1699 (Vasdravellis 1970: 168) mentions the extortions of armatoles who demanded greater payment for their services than they were entitled to collect from villagers, and who exacted illegal transit taxes on travelers. In many instances armatoles became, in effect, licensed governors of mountain areas which the Ottoman authorities did not have the means to police. These arrangements, in some provinces, introduced permanent political institutions and the establishment of families which held captainships through two or three generations and developed alliances through the marriages of daughters with neighboring armatole leaders. A class of licensed Christian militia was created in relations of mutual but shifting dependence on the Turks. They were also involved, on the one hand, in a web of relationships with the local notables of the villages they protected (who in turn had their own internal divisions and their own individual links with Ottoman officials) and, on the other, with the klephts, from whose ranks they originated and whom in so many ways they resembled. An armatole, dismissed, hoped to become a klepht. A klepht by the severity of his depredations hoped to force the Turks to offer him a captainship. A certain level of brigandage was needed to justify the office, yet too high an incidence of disorder removed the captain. It must be stressed that the principal ambition of the great majority of klephts was service as an armatole. This represented for them a partial reinstatement in civil society in a favorable position of influence, pride, and profit. They apparently attached no romantic value to the brigand life, which indeed was dangerous, uncomfortable, and brutal. It was the threat of disorder which had removed them from society and, at a different level, it was the threat of disorder which they hoped would bring them back. The traveler David Urquart reports the flavor of these problems on a visit to a local armatole in Thessaly, one Captain Anastasi. The Captain is speaking:

> "I am very uncertain," he said, "as to my own situation. The Pasha promises me the captainship of the whole country if I can clear it of robbers; but, when I have taken them, will he not demand them from me? Now I can only clear the country by my 'Chatir.' Nothing is easier than to make the bands that now range the mountains, as docile as the men you now see about me; but I can only do so by the faith I would have plighted, and they would have accepted. But if, when I get them into my power, the Pasha demands any of them, either I must fall out with the Pasha, or become an object of contempt to my own people; and I therefore cannot proceed against the robbers till the Pasha gives me his word that he will not require that any of them should be given up. This he will not listen to: and then he orders me to clear the country; and these fools of Primates, notwithstanding all their fine speeches, are just as bad as he is; I have therefore a great mind to turn klepht myself. They are going to send their contributions of silk to Salonica, which amounts in value to 20,000 piastres. They won't allow any of my men to escort it; and they have collected eighty peasants, armed with muskets, no better than sheep. This is an affront; and my men are all mad at the insult that has been put upon them." (Urquart 1839: vol. II, 107)

There is an additional factor to introduce into these remarks on klephts and armatoles. After the Peace of Belgrade in 1739, more use was made of Muslim

Albanians (Skiotis 1975: 316ff.) particularly in garrisons to guard important passes. Albanian irregulars were also found in the ranks of the armatoles but, as Skiotis persuasively argues, there is no evidence, as has sometimes been suggested, that the Ottoman authorities adopted a general policy of replacing Christian armatole captains with Muslim Albanians. Experience had taught the Ottoman Turks that Albanian Muslims were not necessarily more loyal than Christians. Moreover, such a policy would have undermined a security system whose stability depended on a competitive balance between armatoles and klephts which absorbed the energies and ambitions of those men in mountain villages and shepherd encampments who, for reasons we have examined, were disaffected or exiled from their communities, or wished to follow the profession of arms as armatoles (but not as klephts). The mutual surveillance and interchange between armatoles and klephts, on which this balance depended, required that both sides belonged to the same mountain communities. Ottoman difficulties stemmed less from an inherent deficiency in the system of armatoliks than from the general decay of the Ottoman administration, of which one feature was the rapidity with which local Ottoman governors were appointed and dismissed. This practice had developed partly from distrust, partly from the avarice of those in authority in Constantinople who wished to benefit from the bribes and presents which accompanied new appointments. In his turn each new governor made some changes among the captains of armatoles and the commanders of garrisons guarding the important passes.

A consequence of this insecurity of employment in the security forces was that the Pindus mountains were populated with Albanian brigands as well as Greek klephts. It was not uncommon for klephtic bands to include both Christians and Muslims. A document I quoted earlier revealed a Christian and a Muslim sharing the leadership of a large group of brigands. This was not an isolated instance. As late as 1799 a Muslim Captain Bracho led a force of 400 brigands in a devastating attack on Thessaly. Two thirds of his men were Muslim Albanians, one third Christian Greeks. The fact that a man was a Christian or a Muslim was indeed a fundamental element in his identity, but in an ecumenical empire where for centuries the two religions had coexisted (it is true on unequal terms) this had not yet become a boundary for political mobilization and struggle. It would take several years of revolutionary war after 1821 and massacres on a grand scale to effect that change. In the late eighteenth century men, at the local level, could still make alliances across this line. Greek and Albanian mountain people were not very different in dress and manners and their methods of guerilla warfare were almost identical.

Ali Pasha's court at Jannina at the end of the eighteenth century (1787–1822) gave employment, experience, and prestige, to most of the Greek captains who were later to become famous in the War of Independence. He continued to use Greek armatole captains until the end of his period in power. They were at the heart of his security system outside the towns, while the core of the standing army was composed of Albanian contingents (Skiotis 1975: 317). At the same time, however, he ruthlessly hunted down recalcitrant armatoles and their klephtic partners in a supreme effort to establish public order in western Greece and encourage internal trade. By 1805 the mountains were relatively free of unlicensed irregular forces and brigands. A consequence of this was the concentration of klephtic refugees in the Ionian Islands where perhaps for the first time these warriors had some intimation of their common ethnic future as the principal military force in a struggle for political independence.

For the founding mythology of the modern Greek state, which was established in 1830 after the long years of revolutionary war, there was much that was unsatisfactory about the historical truth of klephts and armatoles before 1821 (Apostolakis 1980: Herzfeld 1982). Even for the anonymous composers of klephtic ballads in the eighteenth century, which celebrated the events of the day, there were aspects of reality which were not worthy to be immortalized. Men subscribed to values of honor but in practice they did not always live by them. The ballads therefore adopted the solution of largely ignoring the drama of events and instead concentrated on heroic attitudes and the drama of the soul.

What then can we say about the nature of the hero presented in the ballad material? In the first place he is, like Achilles, an exceptional being with exceptional powers of endurance, not because like that hero he had an immortal mother but because the context and condition of his life makes it possible for him to refine and discipline his sensual appetites, to the point where spirit can achieve what the body alone cannot. A ballad about the famous klepht Nikotsaras has the lines:

> Three days he makes war, three days and three nights
> Without bread, without water, without closing his eyes in sleep,
> Snow they ate, and snow they drank, while they kept on firing,
> And on the fourth day Niko called to his pallikaria,
> Hear me my lads, few but brave,
> Put steel into your heart, and iron into your feet,
> Put aside your guns and draw your swords
> Let us make a charge, and reach Pravi. (Politis 1973: 58)

Although the ballad expresses a sense of obvious pride in the klephts' ability to overcome physical hardship, it is not the case that Nikotsaras endures despite privation, but rather because of it. Fasting enhances the power of the spirit and releases men from their dependence on sleep. In some contexts sleep is associated with death. Among the Sarakatsan shepherds, to fall asleep at a wedding during the three days and nights of drinking and celebration is not merely a question of bad manners, it is seriously inauspicious. Another certain sign of exceptional personal power is the ability to leap. Nikotsaras could clear at a single bound three carts piled with thorns to a height of eight feet. Greeks are not oppressed by a sense of personal sin (Campbell 1964: 324); but there is a clear recognition of original sin, the consequences of which affect everybody. Sin is a weight, *varos*, that literally weighs men down. Only through grace and abstinence are these effects overcome, as in the feats of Nikotsaras. The other expression of endurance which is often represented in the ballads is the ability to resist pain. By all accounts (not only the evidence of ballads) klephts showed no emotion at the torture which habitually accompanied their execution. The famous Katsandonis, still weak from smallpox, was captured in a cave together with his brother. At their execution in Jannina he cried out with pain as they smashed his legs with lead mallets and his brother, who was suffering the same agonies, called out to him not to behave like a woman.

In the ballads, and probably in reality, klephts rarely had access to women. But abstinence from sex in military as in spiritual exercises offered a strength and a purity of spirit which turned away bullets from their mark. One must not misunderstand the significance of these examples of the link between the discipline of sensuality and almost paranormal and prepotent powers. The klepht was not an ascetic. After the exploits made possible by abstinence there were banquets on the Homeric scale with wine and roasted sheep. The movement from abstinence to invincible exploit to celebration precisely parallels a similar sequence of abstinence, receipt of grace, celebration, which occurs at religious festivals. The klepht gave himself positively to life precisely because he knew death was his ever-present companion. He operated with his "soul in his teeth" or, as the Sarakatsani still say, "death is behind our ear."

Thus the mortally wounded klepht wanted to think, even at that moment of approaching death, of celebrating life. As it goes in one song:

> What can I say, my lads, my poor pallikaria,
> the wound is poisonous, the bullet bitter,
> Draw me up and hold me now,
> Bring me sweet wine to drink and
> I will sing songs of grief and sorrow
> that I am not on the high mountains where in the
> deep shade the sheep and the fat rams remain.

As a matter of historical fact the great majority of klephtic leaders about whom we have biographies died violently. In the ballads death is not presented as an extinction of personality. The klepht's fame lives on, in the songs themselves, of course, and in the metaphor of the marble tomb where he can rest with his weapons (weapons whose value must never be dishonored by selling them for money). The tomb has a small window through which the sun shines at dawn and various birds can bring him news from the outside world.

The death of the young pallikari is the central feature of the Greek heroic tradition. It presents the victim with only one fear, the same which faced the

Homeric hero, that of the mutilation and dishonoring of his dead body by the enemy. The nature of klephtic warfare would make it extremely difficult to carry away the body of a fallen comrade, but it was a sacred duty, ideally at least, to cut off his head and prevent that part of his body, which identified him, from suffering exposure and dishonor. So we have in a ballad:

> Like a great tree he was uprooted, like a Cypress he fell,
> In a thin voice he whispered,
> "Where are you my dearly beloved brother,
> Turn back and take me, take my head,
> that it may not fall to Yusuf Arapis and the hunters
> Who will take by body of Jannina, to Ali Pasha the dog." (Politis 1973: 29)

The grief of women and the disorientation of their world is an inevitable consequence of the warrior's death. In the case of Hector, Homer presents this theme through the relationship with his wife Andromache. Generally the klepht was not married, but in any case he lived in a society where the bond between mother and son was closer and more emotionally charged than the more formal and external relationship between a man and his wife. The lament of the mother for her dead son was the bitterest to bear. The model of the Virgin's lament for the dead Christ was a close parallel which had (and has) many expressions in popular tradition. But in the klephtic ballad this dark moment is presented to us from a particular angle.

> They have taken Kitsos, they take him to be hanged.
> A thousand guard him before and two thousand behind
> And behind him goes his mother all in black
> Calling her lamentations,
> Kitso, where are your arms and your empty armor?
> My poor mother, my poor mother, out of your mind,
> You don't weep for my black youth, for my days as a pallikari,
> For my arms and my empty armor you can only cry. (Politis 1973: 27)

The ballad of which these are the final lines is a lament, yet the grief of Kitsos' mother is suppressed under her indignation that her son is no longer free, and shortly will not even be alive, to give his defiance to those who now threaten to dishonor him. In other words she grieves for her son's honor, not for him. And that is how a woman in the heroic tradition of Homeric or klephtic Greece ought to behave. To what extent real klephts approached these attitudes and sentiments it is impossible to know. What, however, the ballads are expressing are the ideal values of honor of the mountain society of villages from which the klephts derived and in which these songs were mainly sung.

In both Homeric and klephtic warrior society men must present themselves as possessing certain qualities and virtues. For on these honor depends. They must be declared, displayed, and boasted of. A man is what he declares he is until his claim is disproved. If the claim is not made, he is assumed to lack what he has not claimed. The use of armor to terrify an opponent in the *Iliad*, the klephtic concern about arms and accoutrements as a visual boast, the power that issues from the Homeric hero's shout (Griffin 1980: 29–30, 38), and the exchanges of insults which habitually preceded action in irregular warfare in Greece in the eighteenth and nineteenth centuries, fit with this way of thinking. In appearing to be an invincible warrior, a man may be accepted at that valuation with the result that his opponents give way. Hector in panic runs away from Achilles. Greek irregulars in the war of independence certainly retreated precipitately when the situation seemed hopeless. And one may infer that klephts did so too, from the importance of the leader's task to encourage them from the rear. The nature of klephtic warfare, with its legitimate emphasis on tactical withdrawals, made the danger of disorder in retreat particularly acute.

In these cultures the fear of failure, the importance of achievement, gave significance to ends before means. Hector boasts over the dying Patroclus, although his death is the work of Apollo who strikes him down from behind. Patroclus is then wounded by the Trojan Euphorbus. Only at the very end does Hector intervene to finish him off. What matters is that Hector does indeed kill Patroclus. In klephtic warfare it is the ambush and the unexpected paralyzing volley which is admired. In the same society, when honor requires a killing in reprisal, the avenger waits behind a rock and shoots his quarry from behind cover. Chivalrous attitudes in this form of warfare would be inappropriate to its purposes and the values which underly it.

It follows that compassion also is a sentiment which is not acceptable in many contexts. Strength in body, followers, and resources is the prerequisite for survival in a warrior society. To be merciful is to be seen to be weak. Among the mountaineers from whom klephts were recruited, manliness is related to the notion of being *varvatos*, well endowed with testicles and the strength that is drawn from them. It contains an idea of ultimate ruthlessness and force that ought, however, to be disciplined to ideal ends if a man is to claim that he is honorable, *philotimos*. In normal practice gratuitous violence is not approved in this society, even among klephts. In transactions of vengeance, sheep theft, or the ransoming of captives, klephts moved cautiously and prudently, taking a life only when they had to. In the *Iliad* Achilles declares that before the death of Patroclus he preferred to spare Trojans when they fell into his hands alive. Nevertheless, in either society, when honor itself is at stake, all such considerations become secondary. We have seen how captives were executed without mercy if a ransom was not delivered. In all cases in the *Iliad* where supplicants beg for their lives in battle, their pleas are rejected. Agamemnon kills the two sons of Antimachus. He hews off the arms and head of one of them and sends his trunk rolling through the battle like a drum of stone. There is a fine line between the man of honor and the creature of frightfulness. Yet when a man appears in life or in legend who is unusually cruel or pitiless, as is the case with Agamemnon and certain of the klephtic captains, his honor is not compromised on that account. In the mountain society of villagers and shepherds the terror felt for such exceptional men was mixed with admiration. Honor, reputation, and family interests could be defended only by those who had the strength. Therefore, the display of physical force, of numbers of kinsmen and followers, of material wealth (particularly in animals) greatly impressed ordinary men, even when the claims and actions based on these resources were taken to excess.

The Homeric hero believed that he could not win renown, or indeed remain alive long enough to win it, without divine assistance. We have already remarked on the frequency of heroes' vows and prayers. The klephtic ballad material does not provide us with similar evidence of religious attitudes. The emphasis is on the klephts' courage, self-sufficiency, and rejection of constraints. We know from other sources, however, that in reality klephts wore crosses and amulets, made vows to saints, were anxious to make confession and take communion at monasteries, in the fashion of the society from which they came. Whether grace comes to a warrior in direct support from above, as in the intervention of the Olympian Gods, or for the klepht in the protection of a saint, or whether it issues from material objects which mediate the power of grace, such as the armor of Achilles or the amulets and crosses which protect klephts from bullets, its receipt is not the mechanical effect of rituals or of contact with ritual objects. It occurs within a relationship between hero and Olympian God, between supplicant and saint; or for the klephts in the case of objects which mediate protective power, an effort of will is required from the warrior in confession, fasting, and sexual abstinence.

In all such matters the hero's world is one of struggle and conflict. His task is a recognized social role, the duty especially of the young warrior, restricted to the class of princes in the *Iliad*, but in the egalitarian world of the mountain villages the duty of every young unmarried man. The understanding that the moment of death is an immutable part of the hero's destiny is one way in which his readiness to give up life for reputation and honor is expressed in the *Iliad* and the ballads. The bullet, or spear thrust, that kills the hero is never accidental. There are no surviving mutilated victims of war; such an outcome would not be a matter for honorable commemoration.

As the representative of his community, or his family, the hero defends its bounds. His actions can be presented as standing for order and stability based on balanced oppositions. Some instances have appeared in this paper: the limited incidence of honor killings which sanction the code of behavior characteristic, until recently, of mountain communities in continental Greece; and at another level the balance between armatoles and klephtic brigands in the Ottoman security system. The threat of disorder, however, is never far from men's calculations in a social system in which the final arbiter is self-help. The story of the *Iliad* is in part a recitation of the consequences of the breakdown of conventions which depend on the recognition of the rights and honor of another; in particular the collapse of the relationship between Agamemnon and Achilles, and of the arrangements for the exchange of bodies for burial between the Greeks and the Trojans. In its simplest form the source of this disorder is that metaphorically, and sometimes literally, two opposed heroes cannot touch one another without the victory of the one and the humiliation of the other.[6]

Notes

1 This essay in an earlier form was delivered as the Sir John Myres lecture at Oxford in 1976. I am grateful to the Humanities Research Centre of the Australian National University for their hospitality and to Dr Roger Just for his critical comments during the period of its revision in 1985.

2 For the Homeric material I have followed the general argument of Redfield's perceptive book.

3 This general account owes much to personal communications from Professor John Koliopoulos. His book, *Listes*, refers to the period after the establishment of the Greek state, but its arguments have influenced my treatment of the klephts. My views have also been influenced by the unpublished M.A. thesis of J. C. Alexander.

4 This is similar to the division of the Homeric landscape. *Agrou ep' eschatien* is the line between the tillable lowland and hill-grazing lands. The former is the properly inhabited world of family life and community, houses, gardens, and cities. The grazing land outside is an ambiguous and dangerous area which young herdsmen share with wild animals, lions, wild boars, and jackals (Redfield 1975: 189).

5 The number probably fluctuated between ten and twenty. See Skiotis 1975: 312.

6 Two important books which deal with aspects of the subject of this essay were published after it had been completed. They are Koliopoulos 1987 and Herzfeld 1985.

References

Adkins, A. W. M. 1960. *Merit and Responsibility*, Oxford

Alexander, J. C. 1966. The klephtic institution of the Morea and its destruction in 1806, M.A. thesis, Columbia University

Apostolakis, Yannis 1950. *The Klephtic Song: Its Spirit and Art* (in Greek), Athens

Beaton, R. 1980. *Folk Poetry of Modern Greece*, Cambridge

Campbell, J. K. 1964. *Honor, Family and Patronage*, Oxford

Griffin, J. 1980. *Homer on Life and Death*, Oxford

Herzfeld, M. 1982. *Ours Once More: Folklore, Ideology, and the Making of Modern Greece*, Austin, Texas

Herzfeld, M. 1985. *The Poetics of Manhood*, Princeton, New Jersey

Koliopoulos, John 1979. *Listes* (in Greek), Athens

Koliopoulos, John 1987. *Brigands with a Cause: Brigandage and Irredentism in Modern Greece, 1821–1912*, Oxford

Politis, A. 1973. *The Demotic Klephtic Song* (in Greek), Athens

Redfield, J. M. 1975. *Nature and Culture in the Iliad*, Chicago

Segal, C. 1971. The theme of the mutilation of the corpse in the Iliad, *Mnemosyne* supplement 17
Skiotis, D. N. 1975. Mountain warriors and the Greek Revolution, in V. J. Parry and M. E. Yapp (eds.), *War, Technology, and Society in the Middle East*, London
Urquart, D. 1839. *The Spirit of the East*, London
Vakalopoulos, A. 1964. *History of Modern Hellenism* (in Greek), vol. 1, Thessalonika
Vasdravellis, I. K. 1954. *The Historical Archives of Veroia and Naousa* (in Greek), Thessalonika
Vasdravellis, I. K. 1970. *Armatoles and Klephts in Macedonia* (in Greek), Thessalonika

8

Name, blood and miracles: the claims to renown in traditional Sicily

MARIA PIA DI BELLA

As is well known, honor looms large in the daily life of Mediterranean peoples[1] and it might therefore provide the basis for detailed comparison between these societies, whose structure is reflected in the form their conceptualization of honor takes. Yet, to analyze what "honor" means, one must first grasp its mode of functioning and understand what the possession of honor implies in terms of its social consequences. In Mediterranean societies, where obligations were traditionally founded on verbal agreements, individual honor and, with all the more reason, collective honor, could alone guarantee the respect and execution of a pact. Thus, every man was invested at his birth with a quota of honor deriving from his family and/ or his lineage, since he was a part of the whole which bequeathed to him, among other things, his "share" of this collective honor which it was his duty to protect and develop as an individual. On the other hand, honor belonged *in solidum* to the family or the lineage as such and was transmitted from one generation to another as a legacy; it could be accreted or depleted by individual members, men or women, through their laudable or shameful behavior.

The elements that constitute the honor of the group and the criteria by which it is granted seem to be, to begin with, the chastity of its women; the purity of its genealogy from which are excluded, in order to preserve its solidarity, the intruding son-in-law and the son of the foreign wife (Farès 1932); the courage of its men on the battlefield and, whenever their point of honor is at stake, in the home community; their ability to defend their women from blemish or from kidnap and to avenge their own group for the murder of one of its members by taking reprisals on the murderer's; the generosity displayed, not only towards their own kin, but towards others on the appropriate occasion and above all towards the poor and the foreign visitor; the sacred duty to provide protection and hospitality for strangers in flight or in distress and for their guests (Pitt-Rivers 1977); the mercy shown towards enemies; the soundness of their judgments; the prestige they win for their group through their actions. The possession of all these qualities establishes the high renown which is the supreme characteristic of honor.[2]

Looking more closely at the mechanism of all these constituents of honor and reducing them to their essentials, one can see that the different visible aspects of acts of honor have their source in the non-visible and that they all stem from two fundamental concepts exclusively responsible for its functioning, namely, Blood and Name; these we can define as the two poles between which the politics of honor are balanced. But honor is also the means of unification in that it connects the feminine "inside" world to the masculine "outside" world. Its most fundamental function, however, is that it connects the past of a society with its future, for honor is the first visible expression of society's awareness of itself in time and of its determination to become involved in history.

Honor: complementarity of Blood and Name

Let us reconsider our definition: the honor of a group depends, ideally, on the complementarity between the two criteria on which it is founded, Blood and Name. Purity of blood, owed essentially to the behavior of women, and renown, due principally to the behavior of men, ensure together their collective honor which is degraded into shame if either Blood or Name is tainted. To restore it, honor must be cleansed, albeit by violent action (see Fig. 8.1).

Honor: unification of the inside and outside worlds

The roles of men and women in the Mediterranean seem at first glance totally separate and opposed; but in both cases, as Bourdieu's (1965) analysis so well shows, whether in the world of the women (the inside world, that of the left hand) or in the world of men (the outside world, that of the right hand), each sex is seen as responsible for its particular sphere, the women for the purity of the family's blood which they transmit and the men for its repute.

The role of women, of primary importance despite the fact that it is exercised back stage, was practically always the same throughout history: to assure the continuity of the group by reproducing it and to safeguard its immaculate genealogy by their own chastity; the woman is thus transformed into a tabernacle of purity, that she has to preserve in her body by remaining a virgin until marriage. Hence, in Sicily, at dawn on the day following her wedding, her closest relatives will pay her a visit while "the nuptial bed is stripped by the mother-in-law herself who establishes with joy and pride, showing the signs to her intimates, that a *virgin* has come to give radiance to her son's house" (Salomone-Marino 1879: 197). After her marriage she has to preserve her purity by her confinement in the part of the house reserved for her. If by chance she has to go out, her attitude must be stamped with a reserve, manifested by her clothing: by the veil in Muslim countries, while in rural Catholic milieux she goes out, if uncovered, wearing her apron to point out the urgent, momentary and obligatory aspect of her errand.

Her role, apparently passive, is to preserve the blood she has inherited, and to hand it down to her offspring in the pure state in which she received it; she must be treated with awe and respect, and her status will always be different in essence from that of men, in that it is based on symbols and not on acts of bravery and achievements. These positive symbols are linked, in opposition, to others with negative connotations, such as menstrual blood, which is regarded as having maleficent magical powers. Quite apart from its power to damage various forms of animal and plant life, Sicilians believe, according to Loria (1907), that women mix it with the bread they offer to the man they want to enamor.

The role of men, as opposed to women, is always played, once they reach adolescence, on front stage, even if it changes during their lifetime according to the responsibilities they assume. When they are young they have to give proof of their courage and their ability to defend their point of honor, to show that they are worthy of their family's name. When they are older, they have to show evidence of their capacity to lead their familial group with the sagacity and prestige of their father. Their role is delicate, for they are constantly pressed into intrepid actions by other members of the family who urge them to add to their collective reputation by a gesture of panache or a demonstration of

Figure 8.1 The complementarity of Blood and Name

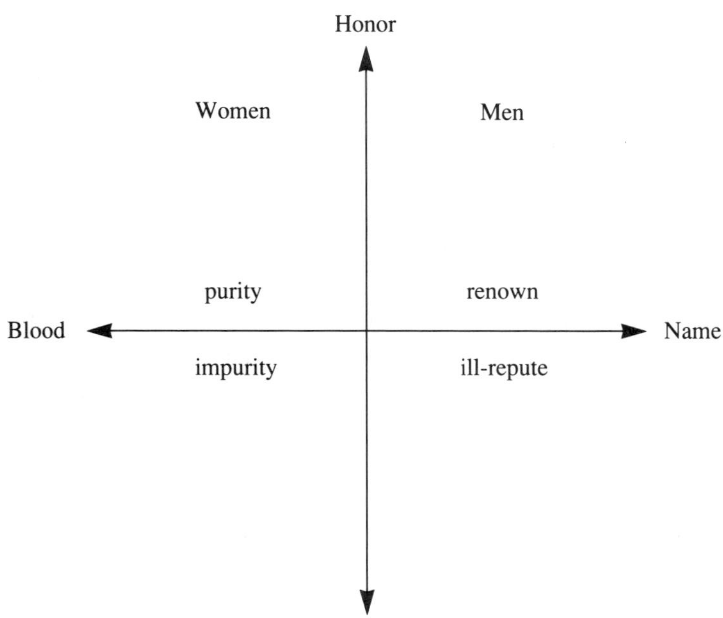

generous intelligence and fine judgment. This continual expectation, by their group and the community at large, incites them to transcend the deeds of their forebears, at the evident risk of committing the mistakes of the over-ambitious. Hence the contests for honor entail a progressive raising of the stakes.

The social expectation for men who belong to less important families is trivial, though they have opportunities, when they are young, to prove their courage and they can, if they seize them and act with address, achieve the highest prestige. For in a society where each man thinks he has honor, either through his personal or familial repute or through the modesty of his women, it is inconceivable that no opportunity at all should be available to a young man if he has courage to prove himself and acquire renown. Young men are therefore tacitly invited to show what they are worth, and, though their errors can be justified by the fact that they are still youngsters (*picciotti*), it is according to their performance at this stage that they will be entitled to enter the elite of the town, or remain forever excluded from it. Since honor is often gained by the youngsters at the expense of older men, conflicts between the generations can become serious, for the desire to destroy an opponent is sometimes so impelling, that to obtain his fall, the very constituents of honor can be ignored (Di Bella 1983). Loss of honor is the only thing the competitors dread, but since they may even increase their honor by depriving others of theirs, the competition can become so frenzied that all forms of affront are possible, and even murder may be envisaged.

Thanks to honor men and women are both committed to the same objectives: to maintain the genealogical purity of the group and to safeguard the renown of the menfolk in the face of the community. The complementarity of the feminine and the masculine role becomes apparent when this balanced unity is destroyed by an offence. There is offence to the Blood when an outsider defiles the chastity of a woman of the group with or without her consent. Once the link between the purity of the Blood and the integrity of the Name is broken, the men, responsible in the eyes of society for their repute, are obliged, in order to efface the insult, to eliminate the woman in question. The problem, easily resolved in the case of a young girl without issue, becomes more complicated if it concerns adultery, for even in the event of the offender's death or departure, the family honor remains blemished whether or not there are illegitimate children. This explains the solidarity Sicilian women display when one of them takes a lover. Salomone-Marino relates that: "the neighbors, the *comari*, the relatives most immediately concerned – those from the husband's family – comfort and help the adulteress in a charitable attempt to resurrect her from disgrace, but only if it is an ephemeral and unique event. They conceal everything from the men working in the fields to avoid murder and the ruin of the family." (1879: 260).

There is an offense to the name of the group when an outrage is committed, either verbally or by some act, against one of its members. The shame entailed is keenly felt by the women who, with words, songs, or actions, encourage their men to restore the family honor. The endless Sicilian or Sardinian vendettas invariably conceal at least one woman who ensures that men take revenge for a murder. When the coffin of the victim is presented to the public, the woman sings, as an *attitadora*, declaiming the death of her brother, father, or husband, announcing that he will soon be avenged. Bannedda Corraine's *attitu* for the death of her brother Carmine sung in 1905 at Orgosolo (Sardinia), is still well known:

> With the death of the honored *compare* our family is fallen,
> with the death of my *compare* the family is brought low,
> my brother was killed by one who was
> the worst of his family.
>
> Mortu 'ompare norattu s'ereu torra a bassu
> Mortu 'ompare meu torra a bassu s'ereu
> Mortu ada a frade meu su pejus de s'ereu. (Cagnetta 1975: 194)

The moral pressure exercised by the women seems to be general to the agonistic societies of the Mediterranean and it is to be noted that they did not confine themselves to prompt action only *after* the offense to the name, they also took it preventively, when there was a danger of offense. To give an example from another field, Arabia: "During the battles, they escorted their menfolk to the fight. Then they struck the tambourines to stimulate their ardor. On occasions, a young one would pick up the flag from the hand of a fallen bearer" (Farès 1932: 123). And, when the women were absent from the battlefield, the men always invoked their sister's name during the hardest moments to recover their courage.

Honor: union of the past with the future

The code of honor, such as it has been described, is the earliest expression of a society's awareness of itself in the past and in the future in that, thanks to its genealogical purity and the exploits of the "fathers," the group acquires a prestige transmitted from forebears which gives it control over its history. It follows that it is more highly esteemed to inherit certain types of past rather than others. This explains the manipulation of genealogies (Peters 1963) and the disappearance from the collective memory of elders who have bequeathed neither glory nor honor and, above all, those who have done the contrary.

Every group displays the same aspirations in so far as it needs a minimal quota of honor to initiate dealings with the neighboring group, who must in their turn have the same quota to make the exchange between them possible. The necessity for all to possess a minimum of honor, to promote a claim to this minimum and to effectuate equivalent exchanges was perhaps at the origin of this "egalitarian" ideology that many observers have noticed among the members of these societies. Honor, then, seems to be present at the institution of every group; it is the substratum on which each one is established, a *sine qua non* of its coming into being, since the constitution of the group is determined by it.

Therefore, people who rationalize their acts and fabricate an image of their society that validates its pretensions to honor are employing a conscious strategy to give themselves a code by which moral, legislative, and executive decisions are taken; as Farès says:

> Arab society owed much to honor: the respect of the totality of its elements, amounting to a code of laws, contributed to the establishment and maintenance of social order. Moreover, thanks to honor, private property, sexual mores and respect for human life were upheld. In effect, protection of the weak was a defense against the violation of property; the concern for the purity of the genealogy put a restraint on sexual licence; the vendetta, finally, established security in the desert (the fear of retaliation obliged the Arab to respect the life of others). In addition, being the first principle of individual and tribal relations, fidelity to one's word was the foundation of the sworn pledge, equivalent to a contract. (Farès 1932: 211)

Familial groups build their economic and political strength on the basis of the honor they have inherited; subsequently this develops according to their economic and political progress. Thus, the necessity for the group's members to preserve their potential of honor and to develop it by themselves and for themselves implies a certain confinement to the group and a constant concern for its interests. Since social advancement is due to honor, every group draws the resources for its further development from its existence as such and the image that it achieves in the eyes of others. The integrity of Blood and Name is essential for its survival, but since the certainty of this integrity can only be guaranteed through a detailed knowledge and a severe control of its members, their range of commitment is limited to their group which provides the only point of reference. The practices of endogamy, widely diffused throughout the Mediterranean, and, in the Islamic area, of marriage with the father's brother's daughter can only be understood in such terms.

The system as we have described it is essentially static; we will never find it occurring in fact in such a coherent form, since it is susceptible to transformations or atrophy; hence we may witness a degradation of the aspect of honor deriving from either Blood or Name, with a corresponding accretion of the other aspect. Thus, as Pitt-Rivers (1977) describes it, the women of the aristocracy do not need to display their modesty and their chastity in order to preserve the honor of their family, the strength of their Name makes it impregnable. In this case honor becomes autonomous, self-sufficient, nourishing itself on its own legend, and insuring the Name's continuity simply by reference to the past. On the other hand, families whose Name is poorer in prestige employ their women's chastity to uphold family honor. Hence, frailty in one aspect can be compensated by strength in another in order to maintain honorable status.

The historical vision of the function of honor fits with the linear conception of time that we find in the Judaic, Christian, and Muslim religions. It will be seen, in the second part of the chapter, that the structure of the notion of honor is reflected in that of grace.

Grace: equilibrium of vow and miracle

The concept of grace is as widely diffused as honor among the peoples of the northern littoral. It is found in the Old but above all in the New Testament; it is also, in varying forms, a central tenet of the Christian faith and it has equivalents in Judaism and Islam.

As we pointed out, Mediterranean societies were formerly based upon the oral pact, which was also, in the form of the covenant, the basis of their relations with God and the saints; the latter were invoked more for their own individual attributes rather than for their capability to intercede with God, especially in Sicily where the protracted absence of the monarch obliterated him from memory and reinforced in the eyes of the population the importance of his representatives. Personal and collective grace guaranteed the respect and the execution of this pact. Thus, collective grace, like collective honor, belonged to the family and was transmitted as we shall see from one generation to the next. Therefore, every member of the family had, at birth, his share of the collective claim to grace, and an opportunity to develop it by his religious fidelity and devotion. He could also destroy it, for himself and his whole family, by his graceless conduct.

The different elements that compose grace, and the criteria by which society recognizes the "gracefulness" of the faithful and of the saints, can be summarized briefly; they are: the request for a favor or a miracle from the saint of the devout person's choice, a request supported, while waiting for it to be granted, by promises, vows, or offerings; the granting of the favor or miracle by the saint; the execution of the vow to thank the saint for the favor or miracle; the many demonstrations made to testify to their receipt; the presence of numerous sanctuaries provoking the saints' generosity in conferring them; the existence in each community of a prestigious patron saint considered as the most miraculous of all saints. The presence of all these elements brings high prestige to the believer, to his group and also to the saints, whose grace is thus made manifest.

Acts of grace, rather like acts of honor, can be reduced to the two institutions of Vow and Miracle and we can see that, for grace to pass, there must be a state of balance between the manifestation of his devotion on the part of the petitioner and a manifestation of benevolence on the part of the saint. This state is ensured by the observance of the Vow on the part of the believer, and the performance of the Miracle on that of the saint. On the other hand, the grace of the group or individual would be converted into disgrace should either the Vow or the Miracle remain unfulfilled. To recover grace the lost equilibrium must be restored (see Fig. 8.2). Grace is also a means of unification because it connects the believers' material world to the saints' spiritual world; and it is also the means of establishing a reputation, sometimes a worldwide one, for the saints and the faithful concerned.

Grace: unification of the above and the below

The roles of the faithful and the saint differ in that the first make vows and the latter grants miracles. Nevertheless, at a closer look, it appears that exchanges take place between them which are possible only if the faithful are considered to be "worthy" of the benefits that the saint bestows. Equivalence between the parties to this transaction is therefore believed to be necessary for obtaining grace.

This equivalence is stressed by publicizing the fact that the believer *obtained* the miracle; in effect, if a miracle concerns a particular organ or member of his body, a silver image or a representation of this part, into which grace has entered, may be hung or fixed to the wall in the chapel dedicated to the saint, to announce the event. Otherwise, the miracle granted may be advertised through the graphical representation of the occasion which gave rise to the request: an accident in which the believer (or a close relative) almost died; a sickness reputed incurable; an encounter with or an attack by dangerous brigands or murderers; a shipwreck. Thus, communicating to the audience the circumstances in which a person obtained grace is a way of making himself known as "worthy" of a miracle, "worthy" of receiving the saint's grace, and therefore "worthy" of renown.

It is sometimes difficult to distinguish a personal request for grace from a collective one, since he who prays for the miracle often makes a personal vow to obtain a collective grace, and while the penitence implied in the vow is naturally assumed by him, the miracle itself may concern somebody else, either a child or a sick adult, and grace may be "received" and accepted by all the members of his family. Nevertheless, requests and grants of grace which are either uniquely personal or on the contrary uniquely collective are also made.

The description of an ancient feast, that of St. Joseph, celebrated annually in Sicily, will be taken to exemplify these statements: on the 19th March, many women celebrate the *Virgineddi* by preparing, throughout the isle, a traditional meal (Pitrè 1913: 187–9; Grisanti 1899–1909: 59–60). This custom, still strictly observed, is usually practiced in exchange for grace received, either recovery from a disease or the guarantee of employment for the head of the family; the feast is offered to at least three poor youngsters or orphans (Giallombardo 1981: 9–15).

Three days before the feast, as Giallombardo observed in Alimena, the women of the group making the vow perform, with the help of other women of the neighborhood, the different tasks necessary for the preparation of the banquet: the gathering and preparation of vegetables, the kneading of flour, the construction of the altar. The day before, they make the mixture in which the vegetables will be fried and the rolls of bread, shaped to suggest parts of St. Joseph's body and/or his occupations: his hands, his stick, his initials, S and G (for Giuseppe), his carpenter's tools. A small altar is erected against the wall between the two narrow tables where the meal is consumed; white veils, pinned with colored balls, Christmas stars, and silver ribbons, cover the wall and hang down on to the altar. On the wall a picture is also fixed representing St. Joseph and the Infant, while on the altar the rolls of bread are disposed according to size. Flowers, wine bottles, oranges, and candles complete the setting. The two tables, covered with white linen cloths, are laid out. Next morning, the priest comes to bless the altar, the tables and the food.

After the blessing, the banquet commences – and it can end only when no one is left knocking at the door. Three or four groups, of about twenty persons each, attend and eat at the table, while the women of the house serve them. When the meal is over, the women sing a fragment of the "Testament of St. Joseph" in which the saint pleads for his own death to avoid the sight of his

Figure 8.2 The complementarity of the Vow and the Miracle

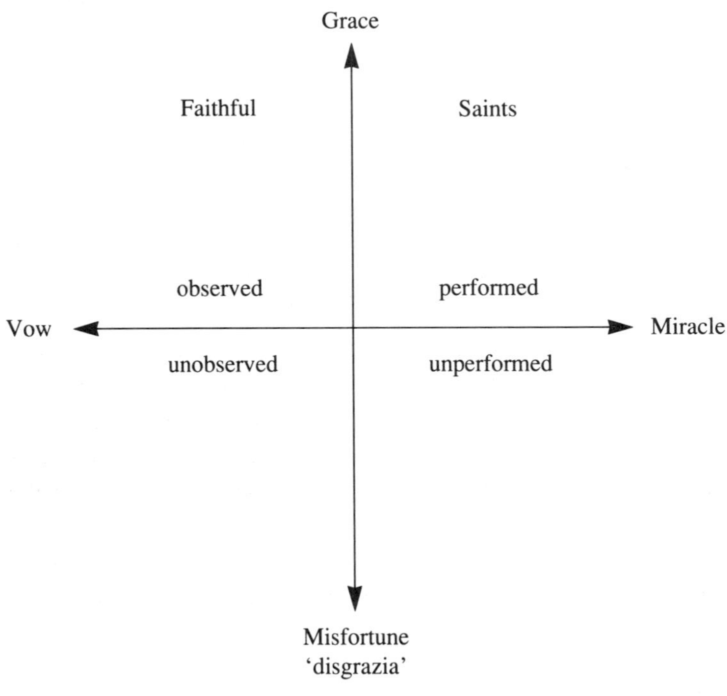

son's forthcoming sufferings. When the guests are about to leave, pieces of the blessed bread are distributed. The final meal is eaten by the hosts, their close friends, and the neighbors who helped to prepare the feast.

This private feast, undertaken entirely by the women of the house, is countered by a public feast celebrated uniquely by the men in honor of St. Joseph. Thus, on the first Sunday of the month of May, the members of the confraternity of the "Company of St. Joseph" go round in small groups through Alimena, with mules to fetch the offerings of wheat which many families make as a promise to the saint for a requested favor or in recognition of one granted. Traditionally, the wheat was distributed to the poor of the village; nowadays its sale covers the expenses of the evening feast. In the late afternoon of the same day, the saint's statue is taken in procession along a traditional route through the village. During the religious ceremony, numerous persons come up with cash offers to the saint; subsequently, their name is inscribed on a list including all devout donors and they receive in exchange a *santuzza*, a small image of the saint.

Throughout the feast in Alimena, the devotee seldom assumes his role single-handed, though the request for a miracle or a favor can be of his personal initiative. When the saint answers a request, the members of the household feel *collectively* privileged, which explains why a vow, contracted for several years, is implemented by others even if the person who made it dies before its completion. The different members of the household express this feeling of collective participation in the vow by evoking the expectations it has awakened in their distinctive spheres: the women in the privacy of the house and the men on the public scene.

The role of the devotee has therefore to be performed in strict adherence to established precept, if the community is to acknowledge him as the effective recipient of the saint's grace or as "worthy" of "incorporating" it into his body, and thus becoming equivalent, in a certain way, to the saint. The saints too, as we shall see, have to conform to the community's expectations; the extent of their adequacy determines the position they will subsequently hold in the community concerned. Their role is to do miracles, but when a saint is "asked" by a member of the faithful to perform one, it means that he has been "chosen" by the community as the best fitted for the role. This choice is generally made either because he seems more capable than other saints at performing the specific miracles requested or because his earthly life illustrates particular incidents favorable to the realization of the miracle. The saint who enjoys the highest esteem in the community is the saint chosen as its patron. Thus, when a believer turns to him with all the humility required to ask for a miracle or favor, his role is "to listen to his prayers and to satisfy his requests" (Salomone-Marino 1879: 175).

Later on, the devotee, proud of his privileged position in the patron saint's eyes, to which the granting of his request attested, hastens to display this relationship. In this way, the number of ex-votos in a church dedicated to a patron saint determines his importance in the community, and the number of sanctuaries or churches dedicated elsewhere to the same saint bears witness to his popularity in the region. Furthermore, the cults, relics, collections, pilgrimages, processions or feasts in honor of a patron saint are a way for the faithful to thank him and to prove to other communities his solicitude towards those he patronizes and his generosity in granting favors and miracles.

The faithful and the saints, bound by the same pact, have complementary roles which link the material world to the spiritual one. Grace is attained by the conjunction of these two roles. This becomes clear when one of the partners concerned breaks the contract between them, either, in the one case, by neglecting to execute the promised vow or, in the other, by failing to produce the requested miracle. The reaction of the wronged partner, aiming to re-establish grace or the lost state of balance, is immediate. There is rupture of the relationship when the devotee, having received the requested favor or miracle, omits to implement his promise of penance, pilgrimage, or feast, or to offer the promised sum of money. The saint then, seeing that their pact has been broken, tries to recall it to him: "in front of the house [of such a person], the float of the saint [*bara*] 'becomes heavier' and the carriers . . . do not move until the Saint receives the gift corresponding to his favors" (Loria 1907: 54). If the devotee persists in neglecting his promise, the saint will either retract the miracle performed or send him a disease or an accident that will require him to invoke his protector anew with the fulfilment of the previous promises and fresh requests for another miracle. Rupture of the relationship also occurs when the saint fails to perform the miracle requested of him. Salomone-Marino (1879: 175–9) describes the different ways in which a personal or a collective request for a miracle is recalled to a recalcitrant saint. The individual reaction consists in rebelling against the saint, by threatening, insulting, and hitting him; by inflicting on him a worse punishment, such as hanging his image head downwards, or shifting it from its privileged place to an inferior one, or locking it in a drawer until the prayer is granted. If the saint repeatedly fails to keep his part of the bargain, the ultimate action is taken: to replace him.

Collective rebellions and punishments on the other hand, occur when saints omit to send the necessary rain to the community. Thus, when St. Angelo, patron of Licata, forgets his duty in this regard, the members of the community take his image to the sea, where they renew their request explicitly: "St. Angelo, either you send rain or you sink." For the same motive, the faithful of Nisseni strip the statue of their venerated St. Michael of its beautiful mantle and cover it with a stained and patched one, then they transport it in great silence to the church of the Capuchins where it remains in penitence until rain comes. When grace is obtained, they remove the Archangel from his place of correction. In drought-afflicted Monterosso, the faithful carry an "Ecce Homo" of papier mâché in procession, but drop it into the horse-pond and leave it in the water for punishment until the rain, the "grace of God" as it is called, falls!

Grace: union of the present with the past

We have been able to see how important it is, for the saint as well as the devotee, to establish a reputation, and how they are linked together in this regard, since each contributes, at different moments, to the making of the public image of the other. A good reputation is engendered by common knowledge of significant events: just as families provide evidence of their daughter-in-law's virginity by hanging the sheets of the nuptial bed on the balcony, so the faithful testify to the miracle they have obtained by hanging in the church an ex-voto that depicts it. It is therefore through the communication of events, usually conducted on a large scale, that individuals validate their claims and develop their potential harvest of esteem.

The miraculous saint and the beneficiary of the miracle are henceforth associated in the collective memory; their renown sanctions their entry into history. But additional devotees will show up, publicizing the blessings they also have received from the miraculous saint, thus adding their name to those already recorded. These newcomers, who try to reproduce during their lifetime the holy actions of earlier known devotees may become, afterwards, as famous as the first ones. Therefore, to reach the state of perfect grace, the actual candidate is usually constrained to refer to the miracles of the past as the constant point of reference for any further claim to spiritual eminence. The condition of access to historical fame is thus a perfect adherence to the tradition established by previous events which govern actions in the present in this manner. Instead of manipulating the past to attain specific future goals, as in the case of the strategies adopted to achieve honor, grace calls for total recognition of its past.

The reciprocal relationship that ties the saint to the devotee and thereby generates grace may also occur at the former's instigation, when saints elect to establish such a tie with persons they decide to favor: innocent children or humble individuals have usually been chosen, from the sixteenth century onwards, to transmit the message the saint wished to reveal to the world.

Thus, if we take the nineteenth-century world-famous examples of the Virgin Mary's apparitions, we notice that they usually happened without any specific request. As the many children to whom the Virgin appeared have related, they were taken by surprise, and they sometimes even mistook her true identity. Maximin Giraud, Melanie Calvat, Bernadette Soubirous or Mariette Beco, aged eleven to fourteen years old, who were privileged to be the witness of these apparitions, respectively in 1846, 1858, 1917, and 1933, all participated in an "unrequested" event, in most cases involving an "unknown" lady.

The locality in which the apparition appears remains sacralized by its occurrence and thereafter attracts the faithful by its potential for miracles. La Salette, Lourdes, Fatima, or Banneux, places considered holy and consecrated since the event, keep their power of evocation, even if no miracle takes place in them any longer for the millions of pilgrims who go there annually in procession: the faithful relive the locality's supernatural past, thus attaching their ordinary existence to that of its privileged witness. The power of attraction of apparitions is essentially due to the fact that they happen *outside* the usual relationship that binds the devotee to his saint, revealing a more direct and spontaneous way of attaining grace. In fact, when we come to examine how the apparitions occur, we find that no prayers are offered, no vows made, to ensure the fulfilment of a request. Grace is given by the saints gratuitously in this context to the innocent persons deserving it.

On the other hand, since the locality is impregnated with the divine presence, the devotees think that they will end by finding there, as the blessed one who was chosen to witness the saint's or Virgin's apparition found, a state of perfect grace. In this case, the miraculous aspect of the devotee's relationship to the saint devolves from his presence on the spot. But only innocent children and subordinate persons, such as the converted Indian who in 1531 saw the Virgin of Guadalupe (later to become the patroness of Mexico), can obtain the privilege of a miracle without requesting it. The humble, who are excluded from the struggle for material possessions or for personal ascendancy, and the young, ignorant of the existence of such struggles, succeed in obtaining what others are willing to acquire at any price, but cannot. Few therefore have a relationship with the saint in which the component of vow is missing, for few receive without giving, and these few are chosen, by an inversion of social status, among the dispossessed.

On the contrary, many have a relationship with the saint in which the component of miracle is lacking, for they are too involved in the mundane aspects of social life to have the necessary qualities to witness a supernatural apparition. For this reason they visit, year after year, the miraculous localities which henceforth embody their desire for a miracle: the visit to the miraculous spot becomes the indispensable substitute for the miracle which failed to happen.

Honor and grace: symmetry of terrestrial and celestial relations

The description of the concepts we are concerned with has stressed the symmetry between the relation which individuals establish with their entourage to elaborate their honor and the one they maintain with the saints to obtain grace. Both these relations were conducted, with equal ardor, by the upper class until around the sixteenth century. After that time, honor began to be claimed on a wider basis and its equilibrium was upset: Name was credited essentially to noble families who were thus able to dispense with its counterweight, purity of genealogy, while Blood, in the form of feminine chastity, was adopted as a criterion of honor by poorer families. Grace in the meantime became accessible, in the form described, to all members of society.

The competition between different individuals or family groups for social supremacy usually took place among persons of the same rank or status, since conceptual equality is the necessary condition for the acceptance of a challenge. Those who, being of a lower social position, were not eligible to enter these contests, turned to the saints hoping to participate, through grace received, in the celestial order and to elevate their status by the same token. Their recourse to relations of grace counterbalanced their inadmissibility to compete for honor, and gradually led to their monopolization of these relations.

The social strategy of the powerful hinges therefore principally on the point of honor while that of their inferiors relates mainly to grace. The first sustains and consolidates the social position of the competing adversaries' Names, which even allows them to erase, if necessary, the stigma of their bastardy.[3] The second, on the other hand, has to be supported by the participants themselves through constant social advertisement in order to establish their reputation. The publicity they give to their relations with the saints makes their name upon the public scene, entitling them to be honored as worthy members of their community and an adjunct to its fame.

Society concedes, to the powerless persons or groups, the opportunity to establish celestial relations with "humanized" saints, replicas of the terrestrial relations they are denied. This search for reputation in the religious sphere leads, as we have already seen, to the social recognition of the person touched by grace, but, at the same time, it enables society to counterbalance possible tentatives of radical change, by offering an alternative to the competition for honor and installing, in this way, a social division everyone seems content with. Terrestrial relations and their celestial counterparts are therefore to be viewed as complementary in maintaining a stable society.

Acknowledgments

I am deeply indebted to the editors of the volume for their help in looking over my English and for their valuable suggestions. I also want to thank Antonino Buttitta, Jonathan Friedman and Pierre Bonte, for taking time to discuss the topic of honor, and particularly the latter, as well as the Délégation Générale à la Recherche Scientifique et Technique for giving me, originally, the chance to work on the subject (1976–8).

Notes

1 For the last twenty years anthropologists of the Mediterranean have had recourse to the concept of honor, in the wake of Pitt-Rivers' and Peristiany's pioneering work, in order to understand the numerous and multifarious societies bordering its shores. My contribution to this endeavor aims to clarify the relation between honor and grace, by observing how each is viewed in itself and how they interact in the Sicilian context.

2 See, for more insight, Bourdieu's (1977) "symbolic capital."

3 Much is made of this point in E. Leroy Ladurie's chapter in this volume.

References

Black-Michaud, J. 1975. *Cohesive Force. Feud in the Mediterranean and the Middle East*, with a foreword by E. L. Peters, Oxford, Blackwell

Bourdieu, P. 1965. The sentiment of honor in Kabyle society, in J. G. Peristiany (ed.), *Honor and Shame. The Values of Mediterranean Society*, London, Weidenfeld and Nicolson; Chicago, Chicago University Press, pp. 191–241

Bourdieu, P. 1977. *Outline of a Theory of Practice*, Cambridge, Cambridge University Press

Cagnetta, F. 1975. *Banditi a Orgosolo*, Florence, Guaraldi

Di Bella, M. P. 1983. Mythe et histoire dans l'élaboration du fait divers: le cas Franca Viola, *Annales ESC*, 4: 827–42

Farès, B. 1932. *L'Honneur chez les Arabes avant l'Islam. Etude de sociologie*, Paris, Maisonneuve.

Giallombardo, F. 1981. *La Festa di San Giuseppe in Sicilia I*, Archivio delle Tradizioni Popolari Siciliane, 5, Palermo

Grisanti, C. 1899–1909. *Folklore di Isnello*, Palermo, Reber (reprinted Palermo, Sellerio, 1981)

Loria, L. 1907. *Il paese delle figure: Caltagirone*, Florence, Il Museo di Etnografia Italiana (reprinted Palermo, Sellerio, 1981)

Peristiany, J. G. (ed.) 1965. *Honor and Shame. The Values of Mediterranean Society*, London, Weidenfeld and Nicolson; Chicago, Chicago University Press

Peters, E. L. 1963. Aspects of rank and status among Muslims in a Lebanese village, in J. Pitt-Rivers (ed.), *Mediterranean Countrymen. Essays in the Social Anthropology of the Mediterranean*, Paris, The Hague, Mouton

Pitrè, G. 1913. *La Famiglia, la casa, la vita del popolo siciliano*, Palermo (reprinted Bologna, Forni, 1980)

Pitt-Rivers, J. 1977. *The Fate of Shechem or the Politics of Sex*, Cambridge, Cambridge University Press

Salomone-Marino, S. 1879. *Costumi ed usanze dei contadini di Sicilia* (reprinted Bologna, Forni, 1970)

Warner, M. 1976. *Alone of All Her Sex*, London, Weidenfeld and Nicolson

9

From the death of men to the peace of God: violence and peace-making in the Rif

RAYMOND JAMOUS

The topics of honor and of *baraka*, or "divine blessing," have been treated in separate studies, but their interrelationship has rarely been investigated. In a previously published monograph (Jamous 1981) I have shown how these two concepts were articulated in the traditional precolonial social structure of the Iqar'iyen, a Berber-speaking group from the eastern Rif (Morocco). In this paper I shall extend the principal arguments of that work to examine certain points, notably: the overarching significance, within honor, of exchange; the relationship between *baraka* and the success of the personal action of certain *shorfa* (sing. *shereef*) or "saints," who are recognized as descendants of the Prophet; and lastly, the relationship between *baraka* and the universal calling of Islam which, in the context of Iqar'iyen ideology, transcends and encompasses society.

For me, honor and *baraka* are not simply cultural entities which govern the life and beliefs of individuals within a given society. Rather, they define the encompassing ideology of Iqar'iyen society and are embedded in its social structure, whose coherence as a whole they render comprehensible. Indeed, it would be difficult to speak of honor without also taking into account its segmentary structure, power relationships, and forms of exchange. Similarly, *baraka* is connected on the one hand to the *shorfa*, who are mediators in matters of honor and on the other hand to the Sultan, the politico-religious head of the Moroccan Muslim community. To discuss the distinctions between society and culture, infrastructure and superstructure, or, more generally, materialist and idealist approaches, does not seem a very fruitful way of identifying and understanding these concepts. Honor and *baraka* are neither the results of economic conditions, nor causes which determine society in all its aspects; they are social facts in the sense which the Durkheimian school gave to the phrase: that is to say, *sui generis* realities which cannot be reduced to a single notion. They constitute that social morality which constrains men and which orientates their interactions within a community.

The honor of men

The Iqar'iyen form an extremely large group of sedentary farmers (a population of nearly 50,000 at the beginning of the twentieth century) who inhabit a tiny area (1,000 km^2), consisting for the most part of deforested, exhausted, and largely infertile mountain. They have constantly been obliged to go outside their own territory to supplement their livelihoods, for their land was far from sufficient to feed them. But if land is a scarcely profitable economic asset, it is the most highly valued social asset. As well as their territory, it constitutes for the Iqar'iyen groups their source of identity. These groups are organized according to a six-level segmentary structure: tribal confederation, tribe, section, territorial community, neighborhood, and finally household. The higher-level divisions are purely territorial units. By contrast, the neighborhood and the household simultaneously delimit territorial units and agnatic groups; patrilinearity and more generally kinship (affinity is a residual element) remain confined within the dominant territorial structure. In this, the case is to be distinguished from that of the Nuer studied by Evans-Pritchard. This hierarchical relationship between territory and kinship whereby the former includes the latter can also be found at the level of those relationships defined by honor.

My analysis of honor can be summarized by two apparently paradoxical propositions. 1. Honor consists in the exercise of authority over domains that are "forbidden" or *haram* (territory in the case of segmentary groups; land and wife for each head of a household), and in the transgression of the integrity of the forbidden domains of others by means of what we shall call exchanges of violence. 2. Exchanges of violence assume different forms: contests of oratory, conspicuous expenditure, killing and physical violence. Their different forms allow a distinction to be drawn between individual honor and the honor of a group. But above all, they simultaneously imply that both men and groups can be distinguished with regard to their honor, and that such distinctions can be annulled and equality restored between the partners of an exchange.

Honor as an exercise of authority over forbidden domains

Honor can be found both at the collective level, each segmentary group being in possession of its own territory, and at the individual level, each head of a household having the right to lay claim to the title of man of honor, *ariaz*, if he possesses a wife and, yet more essentially, land. These two levels are interconnected. Everything which touches on the honor of an individual reflects on the honor of the groups to which he belongs; but the inverse is not necessarily true. Thus any threat against the land owned by the head of a household is also a threat against the territory of his group. By contrast, any attack on the wells owned in common by a group may be confined to the opposition existing between groups. This asymmetry derives from the fact that an individual's land is included within a territory and not the reverse.

Land occupies a central place. It allows each head of a household to be a full member of the different segmentary groups. It characterizes a man of honor and distinguishes him from those who have no land and who are thus placed in the position of dependants. These dependants are of two types: permanent dependants – small groups of musicians, and a few families of Jewish artisans and merchants – who can in no way acquire land, and are obliged to seek protection and shelter from the head of a household; temporary dependants – families of Muslim clients who possess no land or who have lost their land, and, above all, the sons of the head of the household who must wait for their father's death to inherit this asset, so essential for entry into the contest for honor. Such sons are considered "youths" whatever their age may be, even if they are married and have children of their own. They act on behalf of their father, in his name, and are thought to be irresponsible if they try to assert their own personal honor.

The exercise of authority over the domains of the forbidden implies a certain code of conduct with regard to one's dependants and with regard to external attacks. The man of honor must obtain fidelity and obedience from his wife or wives; exemplary conduct and complete premarital chastity from his daughters; obedience without argument in family matters from his sons. In practice it is difficult to maintain harmony within the household. Tensions and conflict always exist between co-wives, between brothers and/or half-brothers, and between a father and sons impatient to enter the contest for honor on their own behalf. But the important thing is to have sufficient authority to keep these conflicts hidden and to prevent them from being revealed in the light of day. If an outsider draws public attention to one of these conflicts or to some blemish within the family, his action will be considered as an infringement on the domain of the forbidden and will call for a response. In the same way the head of a household must protect his domain against all violent aggression: a raid, theft, the murder of a dependant or of a son, an offense against a woman of the household.

But the man of honor cannot limit himself to defensive conduct. He must challenge others by aggressive acts of the same kind. Moreover, each challenge is followed by a counter-challenge and can bring about a fresh series of acts of the same type. For this reason honor exists in the first place in the exchange of violence which, on the social plane, transcends and subsumes individual actions.

Honor as the exchange of violence

I shall use the term "exchange" in the sense which Marcel Mauss gave it in his essay on the gift, and which C. Lefort (1978: 27) has summarized by the formula: "One does not give to receive; one gives so that others may give".

I have called the exchanges in Iqar'iyen society violent because they can all bring about the real or symbolic death of a man of honor. They all involve challenge and counter-challenge. They are transgressions. They assert the existence of the forbidden by violating it, and by questioning the authority of a head of household or group over his own forbidden domains; for the violation does not constitute a basic attack on the system of values. One attacks only what is worth attacking. No one can attack a musician or a Jew without dishonoring himself. Moreover, such people can enter the fields of a man of honor, or go into his house while he is away and see his wife and talk to her, without being deemed to have issued a challenge to honor. The act of transgression can be performed only by a man of honor and not by someone who is marginal. Whenever a man impugns the honor of an individual or of a group he is acting in recognition of its value, and by inviting the person who has been attacked to make a response in kind he confirms his order of values.[1] This response is more than a sanction; it is a counter-challenge by which the worth of the aggressor is recognized and one's own worth affirmed.

Sanctions apply between people of different status and not between equals. If a musician or a Jew commits a crime or a theft, he will be punished by the collectivity without anyone making it into a matter of honor. A comparison is called for here: this sanction which appeals to a sort of judicial authority is marginalized and applies only in rare cases, while the exchange of violence mobilizes and brings into action the whole Iqar'iyen confederation. Our modern societies have inverted the relative weight of these two institutions; they have confirmed the supremacy of a penal law which applies to everyone and which imposes sanctions against their transgressions, and they have made the contest for honor and the exchange of violence marginal, limiting it to such groups as street-corner gangs or even the Mafia.[2]

Of the three forms of exchange, killing and physical violence are the most important, and they color the other two. Contests of oratory amount to provocations and can lead to dishonor, which is symbolic death. Conspicuous expenditure entails the giving of one's resources to be "eaten" by one's numerous guests during ceremonies and festivities. In this context the term "eat" means pillage and devastation. The generous man courts his own ruin and the necessity of being obliged to sell his land and go into exile or become a dependent. Words and goods are no more than tokens for the people who supply them, who thus put themselves at risk and jeopardize their existence as men of honor. And in this area to be a man of honor is simply to be a man.

Without attempting to construct a typology of exchanges, one might suggest that some societies emphasize the exchange of women, which supplies the paradigm for other forms of exchange, notably all the types of prestation and counter-prestation associated with their rituals (Dumont 1957). Other societies, those of Melanesia for example, attach particular value to the things exchanged, like the pearl money of the Are-are in the Solomon Islands (de Coppet 1970), or pigs amongst the Orokaiva of Papua-New Guinea (Iteanu 1984). Amongst the Iqar'iyen it would be difficult to speak of an exchange of women or of goods valued in themselves. All exchanges here tend towards the same ends – real or symbolic death – and all men of honor are alternatively the subject and object of exchange.

Exchanges of physical violence are the most dangerous, but also the most regulated. Every man of honor knows that he risks death when he engages in violence, but he knows that he cannot shrink from that risk if he wishes to assert his worth. As one informant said: "Here a man of honor, an *ariaz*, possesses courage, daring, and a rifle – or he is not a man." Violence takes place according to very well-defined forms which are a function of the type of groups which are in opposition. It is possible to draw a distinction here between the sort of violent exchange which occurs between high-level segmentary groups and that which occurs between patrilineages whatever their segmentary distance may be.

Segmentary exchange of violence

A group decides to organize a raid against its enemy. It steals an animal or grain, avoiding if possible any loss of life. The group which has been attacked replies some time later with a counter-raid of the same sort. Following this exchange, the two groups can decide to return to each other the goods they have stolen, or, if tempers have become heated – which is often the case – to face each other in battle. The battle takes place in an open space according to a well-regulated sequence of events. The young men place themselves in the forefront, vaunting the honor of their group and denigrating that of their adversaries, after which stones are thrown and rifles let off. But everyone is well under cover. Suddenly mediators appear, *shorfa*, who place themselves between the two belligerent parties, soothe their tempers, and convince them to settle their differences peacefully. The stolen goods are returned, and violence averted . . . until next time. The battle takes the form of a mock conflict in which the groups maintain relationships of mutual animosity by means of a form of violence which is highly ritualized and which involves very little loss of life. If someone is killed, the patrilineages intervene and regulate the matter in a manner which I shall describe below. By raids and battle, the segmentary groups do no more than periodically revive an endemic hostility whose causes go far back in time. The whole segmentary structure is thus animated by ritualized acts of violence which are connected in an endless sequence.

Exchanges of violence between patrilineages

In this case no account is taken of the segmentary distance between the groups, and exchanges of violence follow a quite different course. Honor is put at stake at two levels: the honor of the group, and that of the head of household. Every killing or every attempted killing occurs in the wake of a conflict over land or over a woman. While battle places groups in open confrontation and puts the young men, those who are "irresponsible," at the forefront, in the case of violence between patrilineages the initiative is taken by an *ariaz*, a man of honor, who goes to lie in ambush in the mountain and who will kill his adversary by guile. Having committed his murder, he must then stand up and fire his rifle into the air. This gesture is essential, and marks his challenge and, consequently, that of his group. The counter-challenge must be made by a close agnate of the person who has been killed in order to restore the honor of the family and of the patrilineage. If he delays in doing this, a more distant agnate can deprive him of his vengeance and thus gain prestige while the close agnate of the deceased sees his personal honor tarnished. In all cases the counter-challenge must be directed against the original murderer or one of his agnates. The exchange of killings proceeds blow by blow: a dead or wounded man on each side. Once the honor of each group has been restored, public opinion will glorify those men of honor who distinguished themselves in the exchange and will mock those who shrank back or who did not dare to take the risk. Inevitably further exchange will follow, for in this society it is difficult to admit that others are superior in their honor. Those who have demonstrated their bravery will without fail be challenged again, whether by men from their own group, or by others from different lineages anxious to assert their honor. Thus these blow-by-blow exchanges of killings are locked together in an endless repetition which renders uncertain any position of prestige.

The exchange of killing, however, can degenerate into a massacre – and this occurs in very well-defined contexts. As I shall show, some men, more skilful and courageous than others, can come to dominate their patrilineages, becoming "big men," men of power. When, within the same territorial community, two "big men" have thus succeeded in taking control of their respective lineages, the contest for honor becomes transformed into a struggle for power, and it is then that there exists the danger of a massacre. One or both groups are exterminated in the course of a combat of exceptional violence; after which the survivors have no other alternative than to leave their territory with their wives and children. Massacre goes beyond the exchange of violence and beyond the question of honor. It remains very localized and does not spread throughout Iqar'iyen society. The idea of a war involving the whole confederation of tribes is inconceivable and absurd. Let me make an aside here. In an article on Mauss, M. Sahlins sees in exchange a form of political contract designed to avoid the war of all against all.[3] Without going into the details of his argument, it could

Table 9.1. *Modes of the exchange of violence*

Exchange	between segmentary groups	between patrilineages whatever their segmentary distance
Partners	groups	lineages and heads of household
Form	raid	killing
Means of settlement	{ ritualized battle permanent hostility	blow-by-blow infinite series
Result	no distinction with regard to honor	marked distinction above all at the level of heads of household

be claimed that this proposition makes sense only if it is capable of generalization – in which case the Iqar'iyen example contradicts it. Exchange is not here an alternative to violence, but on the contrary includes it. Moreover, the idea of a war of all against all, taken from Hobbes, is anything but validated by a sociological or even mental reality amongst the Iqar'iyen. On the contrary, I would claim that even where violence escapes regulation, it takes on forms which are a function of the encompassing ideology. Massacre is not the war of all against all, but a form of the contest for honor which has miscarried.[4]

The different modes of the exchange of violence are summarised and contrasted in Table 9.1.

Having placed those distinctions of honor which appear between men of honor in the context of patrilineages, we can perceive more clearly the nature of the second paradox mentioned at the beginning of this section.

Exchanges of violence, and prestige and power

It is possible for men of honor to stand out from their agnates in the exchange of violence. But the glory they thus acquire confers on them no title and no particular rights within their patrilineage. It is a different matter when a "big man" or *amghar* appears within this lineage. Patrilineages do not all necessarily possess a "big man" at any particular moment of their existence, much less is he a permanent institution; on the other hand it is not possible to have more than one "big man" within a lineage.

A "big man" has to be more uncompromising than others with regard to his honor. Under no circumstances can he support the misconduct of his wives, daughters, sons, or dependants; he is a past master in the art of public speaking; he will spend without counting the cost and will let no attack on himself go unpunished. He directs all the exchanges of violence within his patrilineage. The challenges issues by his group must come from him, and, before all his agnates, he is responsible for them; challenges directed against any one of his agnates are his concern before anyone else's, and it falls on him to supply the appropriate counter-challenge. Everything happens as if the patrilineage had become one vast domain of the forbidden placed under the authority of the "big man." We shall see that this is only metaphorically the case.

The "big man" arrives at his pre-eminent position after having gained control of the bulk of his patrilineage's land. By various strategies and ruses he contrives to get a number of his agnates badly into his debt and forces them to sell him the majority of their land. This land serves him a double purpose: to support an elevated style of conspicuous expenditure, and to give work to his impoverished agnates who have become his dependants. An ambitious man must prevent rivals within his group from enriching themselves as he has. The success of this dangerous exercise will force his agnates to recognize him as an *amghar*, a "big man," and to place in his hands their fortunes as men of honor.

This strategy pursued by the ambitious man exceeds the rules of honor and of exchange. The future "big man" uses all the expediencies of power to achieve his ends. With the help of his sons and dependants he will make a show of force to check his rivals or any of his agnates who are unwilling to pay their debts. He will use any sort of trick to push his agnates towards expense and ruin. The "big man" is more than a man of standing. He is a man of power for whom other people's honor affords the appropriate means of imposing his will. But this power has limits which honor imposes. Moreover, the constraints of exchange often render the presence of "big men" precarious, and their disappearance always leads to a return to segmentary equality.

The limits of power of the "big man"

A "big man" cannot do without the active support of his agnates to secure the reputation of his patrilineage. All his diplomacy is directed towards making his power acceptable to the very people whom he has impoverished. His agnates will lend him their support only on condition that they retain a little of that land which is so essential to their identity. A man who has been dispossessed of all his land will be tempted to seek revenge or to emigrate. By contrast, a man who has lost only part of his possessions may lend his support to the "big man" in the hope of better days when the "big man" will have disappeared and he, or failing that, his sons, will be able to enter the contest for honor on their own account.

The "big man" cannot, therefore, place under his control all the land of his patrilineage. He remains the first amongst equals and does not change status on achieving his preeminent position. Although he has placed his patrilineage under his own authority, he cannot boast of exercising the same authority over his agnates which a father exercises over his sons. The latter relationship is founded in a form of submission analogous to that which exists between God and the Faithful. It lies at the basis of law and tolerates no transgression. The "big man" cannot lay claim to the authority of this law in his relationship with his agnates. He is only a pseudo-father. Amongst the Iqar'iyen, the relationship between a "big man" and his agnates is analogous to that which exists between an uncle and his agnatic nephew. An uncle is only a pale imitation of a father, and cannot command the same respect. Moreover, he can be expected to deprive one of one's inheritance by trickery just as the "big man" does in a more more developed way. No one will find anything to take exception to if a nephew revolts against his uncle and kills him. Parricide, on the other hand, is considered a sacrilegious act. Similarly, the "big man" is in a position to fear the violent reaction of his agnates, and cannot expect their blind obedience. His power does not, therefore, allow him fundamentally to change his relationship with his agnates, and his position is limited by the requirements of honor.

Constraints on the exchange of violence and the death of a "big man"
The "big man" directs his challenges only against men of honor of the same rank as himself within other patrilineages. Exchanges of violence are effected through the offices of an emissary, but they rarely end in the violent death of either "big man." Rather, they take on the flavor of a mock combat in which one can demonstrate one's force without using it to its full. "Big men" therefore seek to avail themselves of exchanges of violence without having to submit to their constraints. But these constraints are felt in another way.

If a "big man" cannot challenge someone inferior to him in rank, a "little man," this does not mean that a "little man" is denied the possibility of challenging his superior and of thus gaining prestige, despite all the risks which this entails. The "big man" can abstain from responding if the challenge is a minor one, but if someone has been killed, he is obliged to restore the honor of his patrilineage. All the "big man's" craft will be exercised in exporting violence by availing himself of mercenaries. These are men from other lineages prepared to kill for payment. Such action will bring them prestige, but it will also lead the group from which the initial murderer came to take vengeance on them. By this means the "big man" increasingly augments his prestige, but at the same time he becomes the object of an increasing number of challenges. There is no escape from the exchange of violence, and the man of power who is a "big man" cannot halt it; on the contrary, as a privileged target, he risks finding death within it.

The death of a "big man," whether violent or natural, rapidly brings about the disappearance of all inequalities of honor within the patrilineage. The funeral expenses are immense, and can be met only by selling a substantial part of the land he has left. His numerous sons dispute the inheritance. No rule of primogeniture exists; further, the sons refuse to recognize the right of any one of their number to succeed to their father's position, even if their father has named him to do so. The agnates of the deceased will recover their authority as heads of household, and will be able to enter the contest for honor on their own account. Any man who wishes to impose himself on the group must therefore follow once more the same perilous path as the deceased. Amongst the Iqar'iyen, everything is contrived to remind each generation that power is not inherited, but must be acquired by guile and by violence.

Honor and the death of men

In the light of this analysis, it is possible to give a comparative assessment of honor and to specify its particular characteristics in the Rif. In Mediterranean society, honor generally takes two main forms: it characterizes individuals in terms of a code of conduct to which they must conform; it defines the social relations which exist between the different components of society. The dominance accorded to one or the other of these two forms changes the overall significance of honor.

Underlining the particular place of the hero in the Homeric age of Greece, J.-P. Vernant writes: "The ardour and heroic virtue by which they are motivated condemns them in advance to perish in combat in the flower of their youth, and, by the same token, rescues them from the infirmity of old age, from the silence into which the names of the dead sink in common, from the irremediable fate of the forgotten" (Vernant 1982: 12). This glory, won by heroic death, will be extolled, and will make of these heroes "the past" of their group. It defines "a form of honor and of virtue which goes beyond ordinary honor and virtue" (ibid.: 13). J.-P. Vernant then concludes that: "It is not . . . his status or role within the body politic, his position as king, which wins for the hero, as in the case of the Mesopotamian monarch, a death set apart, but the series of exploits which have made him what he is, and, sometimes in open opposition to his community and its recognized leaders, the singularity of his personal destiny" (ibid.: 13). Thus glorious conduct rests with the young hero. It places him on the margin of his community during his lifetime, and, with his heroic death, gives him a universal significance as the "past" of his group.

An analogous situation can be found amongst the Sarakatasni nomads of modern Greece described and analyzed by J. K. Campbell (1964, especially pp. 268–97). Here it is no longer the hero, but the young shepherd, the pallikari, who represents the perfect model of the man of honor. To an extent this characterization derives from the fact that he is a bachelor and abstains from all sexual relationships, but above all from the fact that he is proud, courageous, and will let pass no affront, no insult, without violent response, even if he must lose his life in the process. To be sure, glorious death will destroy him as an individual, but, as J. K. Campbell says, it will leave intact his "persona." The women will weep for him and celebrate him in song, for he is the pride of their family. On the other hand, the married man, that is to say, the pallikari who has not died violently, will be less obdurate, will be willing to lie, and will be more easily prepared to compromise his honor, for he has a family to protect.

In both cases, honor assumes its perfect form in the person of a young man. Courage, contempt for worldly affairs, the search for a glorious death – these are the qualities which make their destinies meaningful, and which make them admired. By contrast, ordinary honor, that of the king in ancient Greece who must take into consideration the interests of the community, that of the head of household amongst the Sarakatsani, is in some sense more socialized, but at the same time less valued.

The Iqar'iyen, however, do not consider that the qualities of youthful and heroic courage are sufficient to define honor. If a young man attempts to distinguish himself by his personal exploits, he will quickly be named as a braggart or as irresponsible. For him, the sole form of honorable conduct is to place his courage and his intelligence at the service of his father. He must learn, in the company of his elders, how to master his words, how to control his impulsiveness. The head of household is usually a relatively elderly man. His claim to honor derives from the fact that he is master over a forbidden domain. This position imposes certain responsibilities on him towards his patrilineage and the other segmentary groups. But there is more to it than this. This position imposes certain responsibilities on him towards his patrilineage and the other segmentary groups. But there is more to it than this. No perfect model of the man of honor exists amongst the Iqar'iyen. The "big man," who is generally cited by way of example, is certainly the person who best fulfils the criteria of honor. But he is a man of power, and those somewhat inglorious actions by which he has been led to ruin his agnates should also be recalled. Besides, honor does not inhere solely in the personal destiny of each individual; more fundamentally it defines the overall nature of Iqar'iyen society. This is more clearly apparent if we summarize our arguments about patrilineages, the segmentary structure, and, more generally, exchanges of violence.

At the lineage level

The appearance of a "big man" in a patrilineage is the result of personal strategy. It modifies the group's internal and external rules of behavior with regard to honor. But the presence of this preeminent individual is a temporary one, and his inevitable disappearance will lead to a restoration of equality of honor within his patrilineage. Moreover, as we have noted, all these groups do not necessarily contain a "big man." The patrilineage is defined first and foremost by its domain of the forbidden: the territory on which it is founded, the land of the heads of household.

At the segmentary level

The power of "big men" remains confined to a low level of the segmentary structure, and only very rarely extends beyond that, without ever throwing into doubt the overall structure of the Iqar'iyen confederation. And what can this mean if not that segmentary equality – which is not the sort of equality known in our modern societies – is more fundamental to the definition of Iqar'iyen identity than the existence of differences of power or prestige. There, land and territory still constitute the referent of honor. But honor appears also in the repetition of exchanges which constantly lead back towards this segmentary equality.

At the level of exchanges of violence

Exchange has sometimes been defined as a means of escaping from individual self-interest by the institution of transactions of an altruistic kind. It is in this sense that one could speak of the pure gift, of sharing, of hospitality, of generosity. Conversely, exchange has been seen as a means of furthering individual enterprise in a different guise (the accumulation of symbolic capital). Indeed, these two approaches are sometimes used by the same authors.[5] But to reduce exchange to egoism or to altruism is still to define it in individual terms and not as a preeminently social value. Amongst the Iqar'iyen, honor is not reducible to an interindividual affair, to rivalries between moral persons attempting to defend their own interests or to acquire prestige. It goes beyond a definition in terms of codes of conduct which refer to corporate groups or to individuals. Honor exists in the endless repetition of exchanges of violence. When a man commits a killing, stands in the open, and fires his rifle in the air to signal his challenge, is he not thus offering himself as the next victim? More generally, prestige acquired in exchange has to be constantly put at risk and will be destroyed by death. The endless repetition of exchanges of violence, as much at the level of patrilineages as at the level of segmentary groups, defines the universal order of Iqar'iyen society. It would be possible to talk here of a relationship of "hierarchical inclusion" in Dumont's sense (1967) between exchanges of violence and individual prestige, between segmentary equality and positions of power, both inhering in the concept of honor.

But a definition of Iqar'iyen society in terms of honor is inadequate. Segmentary opposition in battle never takes place without the intervention of mediators, who are *shorfa* of *baraka*. Exchanges of killing are not always resolved blow by blow, but must at some stage be broken off. Here again the mediation of *shorfa* allows a return to peace. In both cases, honor constantly drives men and groups towards confrontation and death. A return to peace therefore necessitates the intervention of another value, *baraka*, which is also a source of life.

The *baraka* of God

Baraka or "divine blessing" attaches to certain objects or certain beings – such as the Koran, the harvest, the newly born – and renders them beneficial to man. It "falls" on certain men and bestows on them a "force," a "power," which allows them to perform miracles and prodigies. These men are generally, but not exclusively, *shorfa*, descendants of the Prophet. They often constitute lineage groups and live in a tribal setting where they have a particular status. Since the seventeenth century, the Sultans of Morocco have come from a group of *shorfa*, the Alawites. In Iqar'iyen ideology, local *shorfa*, like the Sultan, possess *baraka*. I shall analyze in turn the significance which this divine force assumes within the two categories of descendants of the Prophet, and, more particularly, the relationship which these two categories of descendants have, thanks to their *baraka*, with lay tribes.

"Shorfa" of "baraka"

The *shorfa* of the Iqar'iyen region constitute patrilineages who live in territories sometimes called *zawiya*. Amongst these *shorfa*, *baraka* exist on a double level.

Baraka characterizes the group and its territory. This territory is considered as a domain of the forbidden or a *horm*. The term has the same root as *haram* and comes close to its general meaning. However, in distinction from a *haram* territory which is placed under the authority of a segmentary group, a *horm* territory is placed under the authority of God and forms a sanctuary for whoever seeks refuge in it. *Shorfa* are all called *ahl el baraka*, "family of the divine blessing." This signifies that they too are placed under the protection of God. They may neither be molested, nor participate in exchanges of violence. Peaceful conduct, which is a proof of cowardice amongst the laity, is a virtue amongst the *shorfa*.

Baraka is granted by God to certain individual *shorfa*, and it allows them to perform miracles and to conduct themselves in a particular manner both towards other *shorfa* and towards the laity. These "saints," called *shorfa el baraka*, are essential to the existence of their patrilineages. The first of them, who has set their example, is the founding ancestor of the group. His mausoleum is built within the Izawiya, and it is frequented as much by his descendants as by the laity. The legends which surround his life and installation in the region bear witness to the favors which God has bestowed on him. After his death, there follows a succession of saints, not always from the same line, who occupy successively the preeminent position of *shereef el baraka*. They thus allow their group to maintain its status over time. In the end, one could say that there would not have been *shorfa* lineages were it not for these saints. Here there is an essential difference from the secular groups. For them, only their territorial connection is important in the definition of their identity, the presence or absence of a "big man" being from this point of view a secondary consideration. On the contrary, however, only the presence of the *shereef el baraka* allows *shorfa* lineages to maintain their status and their territory/sanctuary.

The relationship between the *shereef el baraka* and ordinary *shorfa*, or between the same *shereef* and the laity, is one of spiritual subordination analogous to the relationship which exists between God and the faithful, or between a father and his sons. It differs fundamentally from the relationship of segmentary equality which is conceived of as a relationship between brothers. Where *baraka* is involved, there may be no transgressions. It is impossible to attack a *shereef* without incurring divine malediction.

The following list summarizes the differences between secular honor and the *baraka* of the *shorfa*.

Honor	*Baraka*
secular tribes	patrilineages descended from the prophet
domain of *haram*, placed under the authority of men of honor and of the groups to which they belong	domain of *horm*, placed under the authority of God
violent conduct	peaceful conduct
killing and physical violence	supernatural violence sanctioning sacrilegous acts
"big man" presence not essential for the existence and continuity of the lineage	*shereef* of the *baraka* presence necessary for the continuity of the group
dominant model: relationship of segmentary equality between brothers	dominant model: relationship of subordination between a superior (God, *shereef*, father) and an inferior (believer, layman, son)

The relationship between honor and *baraka* becomes concrete in the process of mediation undertaken by the *shereef el baraka* during exchanges of violence. This holy man, assisted by other *shorfa*, will intervene only if he is called on to do so, and never on his own initiative. He is neither judge nor arbiter, and he can impose no solution. Any solution must be accepted by both the parties in conflict.

The solution is easily found between high-level segmentary groups. We should remember here that hostilities assume the form of raid and counter-raid followed by a battle which terminates with the appearance of *shorfa*. These *shorfa* soothe tempers and get negotiations under way. Everyone knows in advance that it will be question of returning the goods which have been stolen. The parleying may take a few hours or a few days, depending on how long it takes to make the young men see reason.

Mediation is more difficult in the case of an exchange of killings between patrilineages. Here it is not a matter of bringing young "irresponsibles" into agreement, but of persuading men of honor. Moreover, there has been the killing of a man, which is the most serious of acts in this society. Mediation occurs only after a first killing. Different reasons drive one or both of the groups to wish an end to violence. All reasons are connected with the fear of being killed and of a massacre. Secret contacts are established with the *shereef* el baraka. He will not agree to intervene until he has established the true intentions of the two parties. Indeed, none of the adversaries can show his fear. The *shereef el baraka* is given to understand by innuendo that they are ready to accept a peaceful solution.

When the agreement of the two parties has been obtained, mediation can become public. The mediation fixes the nature of the *diyith*, "blood wealth," which the killer must pay to the family of his victim. Generally it is a sum of money; sometimes a woman to be given in marriage; and very rarely a servant in the person of the killer. The last two solutions are extremely humiliating for the killer's group, which will accept them only if it is very short of men and afraid of a massacre. Acceptance will eliminate a group definitively, or at least for a very long time, from the contest for honor. The first solution is a more balanced one, and gives rise to hard bargaining. If the victim's group obtains substantial compensation, public opinion will know that it has demonstrated its sense of honor, and that it has made the killer pay dearly for his deed. Alternatively, if it obtains a small amount, its reputation will be yet more tarnished and that of its adversary will go up. Thus even payment of compensation gives rise to competition, and one could say that the *diyith* represents not only the price of murder, but also the price of honor. The *shereef el baraka* has therefore to be a delicate negotiator and able to judge how far the two groups will go in their concessions.

Once a solution has been found and accepted by both parties, reconciliation is sealed by a peace ritual. The killer, with a knife held between his teeth and his hands tied behind his back, sets off towards the territory of his victim's group. He is followed by members of his patrilineage who bring the blood wealth, a sheep, and various condiments. The *shereef el baraka* accompanies them. At the threshold of the victim's territory, a close agnate of the deceased comes to meet them. He takes the knife from the killer's mouth; he unties his hands; and he slits the throat of the sheep in his place. The ritual act is called the *'ar*. From that moment on, the agnates of the deceased can no longer exact vengeance without calling upon themselves divine malediction. The *shereef* blesses the assembly and declares that peace will bring prosperity and riches.

Mediation thus constitutes that crucial moment where one passes from the exchange of violence to the peace of God, and where honor must give way to *baraka*. But it carries with it the term of a return to violence. This double aspect of mediation is as evident in the results of negotiation as in the peace ritual.

In Iqar'iyen ideology, successful mediation is proof that the *shereef* has *baraka*, for only this divine force could make men of honor give way. Thus this value can be asserted only by subordinating the value of honor. But honor, even when forced into the background, continued to play a part. I have underlined the fact that payment of a *diyith* transfers competition from killing itself to the price paid for killing. This competition is far from being simple. For the Iqar'iyen, the only thing worth the same as a dead man is another dead man. Recourse to compensation thus introduces doubt about the honor of the two lineages: the killer's lineage has withdrawn after having made its challenge; the victim's lineage has accepted a pseudo dead man to replace a dead man. In these circumstances, peace can only be a truce, and, in one form or another, the exchange of violence will recur after some lapse of time.

In the peace ritual, the *'ar* constitutes an essential element. The term signifies both sacrifice and shame. The slitting of the sheep's throat in substitution for the killing of the murderer who offers himself as a victim makes reference to Abraham's sacrifice. By this act, the murderer places himself in the position of a symbolic son of the agnates of the deceased, and transforms their domain of the forbidden into a sanctuary, into *horm*, which forbids all vengeance. There are other circumstances – about which I shall not go into detail here – in which *'ar* intervenes with analogous consequences. By sacrificing a sheep before the entrance to the house of an eminent man of honor, by artfully gaining access to his house to touch the grain-mill (a woman's instrument), to shelter beneath his wife's skirt, or to suckle at her breast, any fugitive can thus transform the *haram* of a man of honor into a *horm* sanctuary where he may find refuge. It is a question of submissive rather than aggressive conduct. By *'ar* one switches the value reference: one moves from honor to *baraka*, and the domain of the forbidden changes from the locus of honor to the locus of *baraka*. Here a peculiarity of the domain of the forbidden becomes clearly evident: it is the locus of articulation of two values.

The blessing of the *shereef* of *baraka* at the end of the peace ritual again recalls the particular place of the domain of the forbidden in the peace of God. In the course of his mediation, the *shereef* always promises to both parties that they will have children and a good harvest. Accounts of the procedure regularly add that, from that moment on, the belligerents accept the compromise proposed by the *shereef*. In other words, one cannot halt the process of death except by substituting for it by means of divine force one of life. And again, in this context land and women appear as the locus of honor for which men are willing to die, and as the domain of *baraka*, the source of their society's life and continuity.

'Ar is shame too, as we have mentioned, and can also be interpreted at the

level of honor. The killer has sacrificed his standing as a man of honor to place himself in the humiliating position of a son, of a young "irresponsible," in relationship to those whom he has challenged. The agnates of the person killed have also humiliated themselves by agreeing to accept the killer as a symbolic son rather than taking his life for the crime which he has committed. Sooner or later, then, the exchange of violence will start again in order to wipe out the shame.

In brief, the local social system of the Iqar'iyen involves a necessary separation and hierarchical complementarity between honor and *baraka*, between a relationship of equality and a relationship of subordination, between that which is connected with killing, with the putting to death of men, and that which is connected with sacrifice, with the putting to death of domestic animals. Here is must be stressed that the supremacy of *baraka* is achieved only at the expense of shame at the level of honor, which eventually will involve a return to violence. In other respects, honor is meaningful only in terms of men's passage towards death; *baraka* alone can assure the continuity of society and its regeneration.

The Sultan and "baraka"

The Iqar'iyen do not limit their identity to their own region, but consider themselves to be part of the Moroccan Muslim community. This community is defined foremost by a religious criterion (it is a state which operates according to Islamic regulations) and by the presence at its head of a Sultan who is also called *Emir el Mu'minin*, "Commander of the Faithful." Four tribes, which were artificially created, were placed around the two capitals, Fez and Marrakesh, and supply the personnel of the army, the *jaysh*, and of the administration, the *makhzen*, of the Sultan. If the Sultan used to control directly through his administration the main towns and certain areas of the plains, it was nevertheless a different matter for the greater part of the countryside. There, the Sultan was content to choose from amongst the local dignitaries someone who could represent him within a tribe, and whose principal task was the collection of taxes and the repression of any rebellion.

In order to define the status of the Sultan and his place within the Moroccan community, we must see how violence and *baraka* are connected in accession to the throne. We shall follow the account given by the Iqar'iyen.

Accession to the throne.

No precise rules concerning succession to the throne exist. A reigning sovereign can designate his son, his brother, or a cousin to be his successor. But his wishes will never be followed. In every case each new Sultan has to take his throne by force in the following way.

The death of a reigning sovereign brings anarchy to the country. The tribes pillage each other, no longer obey the representatives of the *makhzen*, organize devastating raids against the towns, and attack the army. Brigands infest the roads, and insecurity is such that no one dares any longer venture outside his own house. Each Alawite candidate to the succession gathers together his supporters in the army and the administration. Conflicts and intrigues develop. Finally, the pretender who has secured the support of a strong section of the army will be the victor. He must exile his rivals (but never kill them, for they are *shorfa*) in the Tafilelt, the cradle of the dynasty, and severely punish their supporters. But that is not enough. To be Sultan, he must pacify the country by organizing punitive expeditions, *harka*, against the rebel tribes. What is said is that he "eats" them. His army ravages, burns and pillages the land, houses and territories of these tribes, and kills many of the rebels whose heads will be conveyed to the capital where they will be pickled in salt and exposed on the entrance gates of the town. The pretender then returns to his capital and is proclaimed Sultan. From all the towns and all the tribes delegations bring him gifts to indicate their submission to his authority.

Baraka, it is said, is what has permitted this new Sultan to accede to the throne by ridding the country of anarchy and reforging the unity of the Moroccan community around his own person. *Baraka* will fall on this community bringing fecundity to the women and fertility to the earth, and so riches and prosperity to the whole country.

This representation of succession to the throne is significant in the two following areas.

Violence and "baraka"

At the local level, one can be a "big man" or a *shereef* of *baraka*, but never both at once. On the other hand, at the inclusive level of Moroccan society, the Sultan is the only *shereef* who can, and indeed must, employ physical violence to secure his power. Such violence is not perceived as being a manifestation of his honor, but as proof of his *baraka*. It is God who has made him triumph over his rivals and succeed against the rebels. Here, as in mediation, defeat is the proof not of human weakness, but of a divine sanction. For this reason the sovereign must exile his rivals and punish their supporters.

If violence has brought a *shereef* to the throne, further violence, that of a powerful rival, can remove him from it. A pretender may lay claim to a *baraka* superior to that of the sovereign, foster tribal rebellion, and put together an army to seize power. This is a constant danger for the reigning Sultan, but it becomes truly menacing during the troubles and political instability which follow famine, natural disaster, or foreign military intervention. Then the Sultan must travel frequently into the country areas to reassert his *baraka*. There is thus at the inclusive level, as well as at the local level, a repetitive movement going from violence to *baraka*. But while at the local level this movement expresses the hierarchical complementarity which exists between *shereef* and men of honor, at the inclusive level it defines the status of the Sultan in his relationship with the Moroccan community.

The Sultan and the tribes

The anarchy which follows the death of a sovereign appears as a reversal of the structural order, and makes explicit the nature of that order. The tribes are no longer capable of establishing internal or external relationships except by recourse to a violence which is carried to its paroxysm in their desire for domination. The presence of the Sultan makes possible the integration of the tribes within a transcendent order defined by reference to divine Law. This order is revealed in the movement which leads the Sultan to kill the members of those tribes which have rebelled, and, once on the throne, to bless the women and the land with his *baraka*. An Iqar'iyen myth perfectly illustrates my argument. Summarized, it goes as follows.

The Sultan Moulay Yacoub abdicates his throne in favor of his son, Sultan *el Kohl*, "the Black." He travels throughout the country and arrives in the land of the Iqar'iyen where the men of one territorial community force him to dance. He then appeals to his son to come and avenge him of the sacrilegious act which has been committed on his person. The Black Sultan attacks the Iqar'iyen and, by means of trickery, kills all the men and male children of the region, sparing the women and their daughters. He then returns to his capital and tells his subjects to go and populate the territory of the Iqar'iyen and to marry their widows and daughters. The account concludes in the following manner: it is for this reason that we Iqar'iyen are the children of women. In another version of the myth, informants are even more explicit: the women pass on the land of their husbands who have been killed by the Sultan to their sons, born of their union with the men who have come from the interior.

I should emphasize the analogy of this account with the facts of accession to the throne. In the first part, the sacrilegious act committed on the ex-Sultan becomes a serious attack on the authority of his son, the Black Sultan, in the same way as the anarchy and rebellion which follow the death of a sovereign threaten his successor. The Sultan then comes and kills the Iqar'iyen as does every pretender who "eats" the country of the rebels. Finally, after having killed the Iqar'iyen, the Black Sultan brings fertility to the women of the region by means of his men, just as the new sovereign brings it with his *baraka*. But the end of the account contains an extra detail which is fundamental. Despite the massacre of the men of honor, the identity and continuity of the Iqar'iyen is doubly assured by the women and by the land, since the imported husbands are reduced to the role of genitors, substitutes for the sovereign's *baraka*. In proclaiming themselves the children of women, and, through women, the children of the land, the Iqar'iyen are affirming that, at the inclusive level, they are the symbolic sons of the Sultan and, at the local level, that they are the successors of men killed by that same Sultan.

Further similarities are to be found in the marriage ceremony. Here, men of honor must hide themselves from their sons who turn on them in derision. If they continue nevertheless to show themselves, they will be reminded of the precariousness of their existence and they will be told of their coming death which is simulated in ritual. This comes about when the groom, who plays the role of a new Sultan, beleaguers by "force" the house of his father, asserts his divinely derived authority, and symbolically makes fertile his young bride by the power of his *baraka*. This scene enacts the murder of the men of honor by divine force, the condition on which the domain of the forbidden assures, as the domain of *baraka*, generational succession. One cannot therefore speak here of patrilinearity in the simple sense generally given the term. When the Iqar'iyen say: "We are the children of women," it is not to point to a matrilineal tendency, which does not exist, but to indicate that the succession of generations within the agnatic line is rendered possible by the ritual transformation of the domain of the forbidden into the domain of *baraka* made fertile by the Sultan.

In brief, the Moroccan community and the tribal communities, far from being antithetical as some writers have asserted, are articulated with each other within an overall structure by the mediation of the Sultan's *baraka*. His *baraka*, which brings into close association the power of death and the power of life, comprehends local honor and assures its perpetuity.

The Sultan and the "shorfa" of "baraka"

The Sultan and *shorfa* mediators both possess this divine force called *baraka*. They occupy, however, different statuses. A mythical account will show us the relationship which obtains between them.

Sidi Abdslam Ben Saleh (the founding ancestor of a *shorfa* patrilineage of the region) arrives from Algeria in the country of the Iqar'iyen at the moment when the Sultan Moulay Ismail (the first sovereign of the Alawite dynasty) is making war against the Spanish to retake Melilla, an enclave situated in the eastern Rif. He goes to see the sovereign and asks him to give him the plain of Bu-Arg (situated at the foot of the mountain of the Iqar'iyen's country) for him and his family. At this period, the plain is thickly forested and inhabited by wild animals. The Sultan violently refuses his request, for, it is said, he distrusts him. After a moment, he asks him, "Why do you want this uncultivable forest?" Sidi Abdslam makes no reply, but raises his hand over his mouth and brow. In an instant the forest burns, revealing a rich and fertile plain. Moulay Ismail recoils in fear before this miracle. He grants the plain to the *shereef*, exempting him and his descendants from all taxes on this vast territory. Moreover, he gives him the right to bear the title of *shereef*.

In this account, Sidi Abdslam lays claim to a plain which, at the time, was situated on the margin of the Iqar'iyen's territory and which belonged to the Sultan. By addressing himself to the Sultan, he shows his intention of becoming his subject. The sovereign refuses him, and asserts his authority towards this *shereef* whom he considers not as a subject, but as a potential rival.

The second part of the account resolves this conflict and inaugurates a new relationship between the two leaders. The *shereef* demonstrates the power of his *baraka* before which the Sultan is obliged to yield. He receives his territory, and becomes the subject of the sovereign, whom he confirms in his position. Thus the miracle and its result allow their relationship to be established at two levels: at the spiritual level, the supernatural violence of the *shereef* of *baraka* is proof of his hierarchical superiority over the Sultan in the divine order, while at the temporal level the relationship is inverted and the *shereef* submits to the Sultan whose subject he becomes. If the *shereef* obtains from the Sultan the right to bear his title of *shereef*, and to possess his sanctuary, this *horm* territory at the edge of the tribal world, the Sultan also sees himself legitimized in his status by the *shereef*, his superior at the level of *baraka*.

Thus the Sultan, who combines in his actions both violence and *baraka* in order to restore the order of divine Law to the Moroccan community, on the one hand assures the perpetuation of local honor, and on the other hand establishes the *shorfa* mediators in their sanctuaries and in the local status.

But the *baraka* of the *shereef* is not in itself localized. The saint's supernatural violence, although established within a particular region, extends beyond that region and even subordinates the sovereign and so the whole Moroccan community. This aspect of the *shereef* can be observed in other manifestations of his *baraka*. The examples are numerous: he can pass through fire unharmed; he drinks wine and his divine power transforms it into pure water; he can change stones into gold; he has the gift of changing himself into a bird or an aquatic animal; he can see what is happening from afar and foresee future events; he can transport himself in the blink of an eye to Mecca, or make Mecca come to him; his death is only an apparent one, and men build him a mausoleum where he continues to show himself through his miracles.

Thus this divine force can act on the elements, confound the dimensions of time and space – notably the perception of distance, the cycle of life and death, and the distribution of the Muslim universe into center (Mecca) and periphery – and finally brave the outstanding pollution of religious prohibition. All these qualities confirm that in Iqar'iyen ideology the *baraka* of the *shereef* escapes any classification and any rule, while all the time being integral to the particular nature of a society. It makes reference to the God of Islam, and is a universal value which transcends and encompasses any region and any society.

If we now characterize the two values of honor and *baraka*, it can be seen that they have different fields of application. Honor is the value which transcends and encompasses individual assertion. It inheres in the totality of a particular society, the confederation of Iqar'iyen tribes. It is characterized by what it includes, segmentary equality and the exchange of violence, and by what it excludes, musicians, Jews and other dependants who are not integrated into the territorial structure. *Baraka* is defined precisely by its power of inclusion. It gives to the tribal community and to the Moroccan community their religious dimension. But it is precisely because it goes beyond any community and finds its origin in a universal transcendent value that it is able to establish the two communities in perpetuity.

The close association between *baraka* and personal success leads me to compare this value with the concept of chance which, in our language and our society, is its apparent translation.

"Baraka" and success

Recognition of *baraka* implies that the *shereef* has succeeded in his mediation and that the pretender to the throne has been able to impose his authority on his rivals and on the rebels. In both cases, failure is the sign that one does not possess this divine force. Thus it is neither a ritual nor a rule of succession which determines accession to status, but personal success which confirms the *baraka* of both.

A comment is called for here. If *baraka* implies success, not all success is conceptualized in terms of *baraka*. The "big man" is someone who asserts himself through his individual actions. The risk – for there is always a risk in the game of power – is that he might fail and that a rival might take his place. A "big man" is thus a man of standing and of power who has succeeded. But he always remains short of *baraka* as we have shown it.

Comparison between the "big man" and the two bearers of *baraka* can be pushed further. The "big man" is part of the system of honor which constantly tends to annul the effects of all individual success. In this context, the power of this pre-eminent man is a residual element in the sense that it does not provide the structuring principle of Iqar'iyen society, but remains incorporated within that of segmentary equality. By contrast, it is impossible to conceive of *baraka* at the local level or the inclusive level within the presence of the *shereef* and of the Sultan. These leaders are responsible for their *baraka* before men, but at the same time they are the passive agents of a power which extends beyond them, a power which God has placed in them, but of which He can deprive any of them. It is thus by their personal actions that they must continually bear witness to the *baraka* present in them.

"Baraka" and luck

The connection between *baraka* and success can be made more precise. In his work on the Moroccan High Atlas, Gellner has quite properly seen that the *shereef* cannot have his *baraka* recognized without succeeding in his mediation – and this success depends on the good will of the parties in conflict. He concludes from this that in this context the *vox Dei* is in fact the *vox populi* (Gellner 1969: 151). Without being wrong, this analysis is clearly incomplete. The Iqar'iyen know well enough that the success of mediation depends largely on the patrilineages engaged in the exchange of violence. But there is a moment of extreme tension where the desire to make peace and the desire to pursue vengeance come into conflict and can make the negotiations miscarry. Complete success cannot come about without that divine spark which exists in the *shereef* of *baraka* to which the Iqar'iyen submit. *Baraka*, for all that it characterizes the status of the descendants of the Prophet, necessarily inheres in women and land as the power of life which gives them their fecundity and fertility.[6] Let us remember that each newly born child and each harvest are *baraka*. And this is not the result of some sort of fatalism. The care accorded to women and land is central to the preoccupations of men. But the latter know that this vital domain is at the mercy of imponderables which are beyond their control, and that their efforts are insufficient to render fertile their women and their land. One could therefore say that the Iqar'iyen possess a notion which borders on that of luck. Moreover, dictionaries and popular usage in France often translate the word *baraka* by this term. But the similarity between the two terms should not mask their difference.

For us, luck is mysterious. It belongs to the domain of the unknown and the unforeseeable, and seems linked with chance. To be sure, scientific knowledge can attempt to reduce the inexplicable or to replace the law of causality by that of probability which takes account of chance, and which specifies the degree of incertitude in prediction. But no science nor any society can pretend to disregard the notion of luck or entirely to eliminate it from its conceptualizations. The mystery of luck remains complete, and one cannot explain it by laws of nature. Whether it is conceived of as luck or as *baraka* is not a question of scientific truth, but of the ideology of the society which so conceives it. In our modern conception which has separated ideas and values, chance is a residual element in a universe of knowledge. By contrast, for the Iqar'iyen, who do not separate values from ideas and action, *baraka* is a source, God, and appears as an encompassing value which transcends social rules and relationships at the same time as it gives them their specificity. It is embodied in those statuses where personal success and divine force are linked, *shorfa* and Sultan; statuses which define the two levels, local and inclusive, of society. It inheres in land and in women, the domain of the forbidden for which each man is responsible. This domain of the forbidden is central to the conduct of honor, but also to that necessary conduct of submission towards *baraka* which perpetuates the identity of society.

Notes

1 C. Lefort (1978: 27) emphasizes the same point about the gift: "To give is just as much a matter of putting someone else under an obligation as of putting oneself under an obligation by accepting that the other will make a return."

2 One cannot, therefore, analyze these two modern examples of honor without taking account of their place within society as a whole.

3 M. Sahlins has written on the subject, "As against war, exchange," and, bringing together Mauss and Hobbes, attributes to them very similar political philosophies, asserting that "The gift is the primitive way of achieving the peace that in civil society is secured by the State" (Sahlins 1972a).

4 The idea of a war of all against all is a modern one, and is based on the logical and ontological primacy of the individual (cf. Dumont 1965, especially pp. 38–42). In this context, society, whether it takes the form of Hobbes' Leviathan-State or of Sahlins' primitive exchange, will be limited in its political aspect and will have no function other than to restrain the destructive tendencies of the state of nature, of individualism. This proposition does not amount to a universal truth, and can in no way be applied to Iqar'iyen society. As this article shows, the social rule determines and gives meaning to individual action, and not the opposite; the rules of the exchange of violence lead to an understanding of a massacre, and not the reverse.

5 Cf. M. Sahlins 1972b. The author distinguishes between three forms of reciprocity, of which the first two, generalized reciprocity and balanced reciprocity, are of interest to us here, for the third, negative reciprocity, does not partake of exchange. Here is how he characterizes the two forms of reciprocity: "'generalized reciprocity' refers to transactions that are putatively altruistic on the line of assistance given and, if possible and necessary, assistance returned" (pp. 193–4). "Balanced reciprocity is less 'personal' than generalized reciprocity. From our own vantage-point it is 'more economic.' The parties confront each other as distinct economic and social interest" (p. 195).

6 It is interesting to note that the Iqar'iyen dedicate a tenth part of their harvest to the *shorfa* and to the Sultan in return for the *baraka* which they have received.

References

Campbell, J. K. 1964. *Honor, Family and Patronage*, Oxford, Clarendon Press

Coppet, D. de 1970. 1, 4, 8, 9, 7. La monnaie: présence des morts et mesure du temps, *L'Homme*, 10, 17–39

Dumont, L. 1957. *Hierarchy and Marriage Alliance in South Indian Kinship*, Occasional Papers of the Royal Anthropological Institute 12, London

Dumont, L. 1965. The modern conception of the individual. Notes on its genesis and concomitant institutions, *Contributions to Indian Sociology*, 13–61

Dumont, L. 1967. *Homo Hierarchicus*, Paris, Gallimard

Gellner, E. 1969. *Saints of the Atlas*, Chicago, University of Chicago Press

Iteanu, A. 1984. *La Ronde des Echanges. De la circulation au valeurs chez les Orokaiva*, Cambridge, Cambridge University Press; Paris, Maison des Sciences de l'Homme

Jamous, R. 1981. *Honneur et Baraka, les structures sociales traditionelles dans le Rif*, Cambridge, Cambridge University Press; Paris, Maison des Sciences de l'Homme

Lefort, C. 1978. L'échange et la lutte des hommes, in *Les Formes de l'histoire, essai d'anthropologie politique*, Paris, Gallimard

Mauss, M. 1990. *The Gift: the form and reason for exchange in archaic societies*, trans. W. D. Halls, London, Routledge

Sahlins, M. 1972a. The spirit of the gift, in *Stone Age Economics*, London, Tavistock Publications, pp. 168–83

Sahlins, M. 1972b. On the sociology of exchange, in *Stone Age Economics*, London, Tavistock Publications, pp. 185–275

Vernant, J.-P. 1982. Introduction, in G. Gnoli and J.-P. Vernant (eds.), *La Mort, les morts dans les sociétés anciennes*, Cambridge, Cambridge University Press

10

Indarra: some reflections on a Basque concept

SANDRA OTT

What do the wind, sexual potency, a pelote player, and a Basque house have in common with the Immaculate Conception? *Indarra*, the divine "force" that came upon Mary to create the Son of God, the "force" that fills sails, engenders children, puts strength in the human body, the jural "power" which, when legitimately acquired, maintains order in the house. To understand what *indarra* is and why it has a conceptual significance for Mediterranean anthropology, we must look beyond the wide semantic range of the concept to see what relations it informs in Basque society, and how, like grace, it combines secular and physical realms with spiritual ones. The aim of this paper is to do so.

In the first part I provide an introduction to the problem of translating and interpreting *indarra*. To give the concept a firm ethnographic base I describe briefly the French Basque community of Sainte-Engrâce[1] and consider *indarra* within the context of the house. In the second part I show in addition how *indarra* is connected with power and equality in three sets of relationships: first, in the relationship between the person who inherits the house and his or her siblings; second, that between elder and younger heads of household; and, third, between husband and wife. The relations of spiritual and secular subordination which obtain between priest and laymen are examined in the third part; and, finally, I consider some of the other ways in which the proper acquisition and use of *indarra* are concerned with legitimacy, control, and socialized behavior.

A word common to all the dialects of Basque, *indarra* has a range of seemingly disparate meanings.[2] Among other things, *indarra* denotes physical strength, force, power, authority, influence, abundance, energy, efficacy, life force, as well as a young shoot. In a manner similar to *baraka* (cf. Bourdieu 1972a; this volume, chapter 4; Gellner 1969; Jamous 1981; Westermarck 1926), *indarra* is sometimes a vital quality that inheres in things or beings; sometimes it is a quality of relationship; at other times a force that acts upon the natural world. Certain things *are indarra*, e.g. gusts of wind, the fruits of the earth, the spirits that reside in Heaven.[3] An example is provided in Job 17:1, when Job is made to say "my breath/spirit/life (*indarra*) is corrupt, my days are extinct." But more commonly things *have indarra*, of which there may or may not be tangible manifestation.

It appears that not everything which has or is *indarra* can be fitted into a single analytical framework, though proper procurement and employment of *indarra* do tend to be concerned with legitimacy and control in social relations, as well as in reproduction. In the Basque community of Sainte-Engrâce, the voice of the church singer had *indarra* (when this office was filled); the moon has *indarra*, in that it waxes and wanes during the lunar cycle. In the animal and vegetable kingdoms *indarra* is what makes all species grow; it makes pigs fat and causes beans to sprout. In a manner reminiscent of *baraka*, beans and other garden produce begin to shoot as "they take *indarra* from the earth." During this delicate stage the young plant must not be picked lest the entire crop should lose its *indarra*. Semen, rennet, and yeast – all life-giving substances in Saint-Engrâce – also have this vital quality. As the life force in humans and animals, *indarra* is thought to be weakest in spring and in autumn; for in spring the blood has to renew and to strengthen its *indarra*. In autumn the *indarra* of blood "is weak and old like sap in trees in November." The river that flows through the community is said to have *indarra*; and in another part of the Basque Country a mountain, the Gorbea, also has this special "force." In both men and women *indarra* as physical strength is highly valued by Basques; and the *indarra* of men is publicly tested in weight-lifting and other festivals that are a prominent feature of Basque festivals. In Sainte-Engrâce it is said that a mad man has three times more *indarra* than a sane one. In some villages of Vizcaya and Guipúzcoa it was once customary for people to sponsor jointly masses for the souls of their dead. According to Azkue (1935: 182) they preferred to give the sacristan coins with which they had already sponsored a mass; for they thought that by circulating the same coins the *indarra* of their offering would be increased.[4]

Although the Basques of Sainte-Engrâce do take an intellectual interest in certain concepts that are central to their society and that very nearly qualify as total social facts, they do not talk about *indarra* readily or easily.[5] While people have more to say about *indarra* when they describe its visible effects, e.g., the disfigurement of a man's face after he laughed in church, they are content to treat *indarra*, in certain contexts, as an intangible essence.

In Sainte-Engrâce *indarra* often combines secular and physical spheres with spiritual, supernatural ones. This curious characteristic of the concept – as well as its wide semantic range – sets it apart from two other Basque words with which *indarra* may be instructively contrasted: *azkarr*, which also denotes physical strength and is applied to substances that have a strong taste; and *zuzena*, which conveys the meanings of right, power, authority, and influence. *Zuzen* is a variant form of *xüxen(a)*, a term commonly employed to describe someone who is just, equitable, honest, and conscientious. To be *azkarr* a person must be physically strong; but *azkarra* (as a substantive) lacks the volition of *indarra*, which combines tremendous stamina, determination, and energy with great strength. Few people are judged to have *indarra*, and those who do are much admired. As I show in my analysis of the house, the relationship between *indarra* and *zuzena* is more complex. Though there is a certain overlap of meaning with *zuzena*, *indarra* is not used interchangeably with it.

Sainte-Engrâce is located in the southeastern corner of Soule in the French Basque country. In 1979 the commune had a population of 376 which was divided into two separate communities, upper and lower. Together these occupy a mountain valley some 11 km in length. Outside the household the two most important institutions in Sainte-Engrâce society are "first neighbors" (in which each house has a permanent, prescribed relationship with three others) and the shepherding, cheese-making syndicates known as *olhak* (sing. *olha*) (cf. Ott 1980, 1981).

The people of Sainte-Engrâce speak Basque as their daily language; they organize their lives around their sheep and small farms; and they are Roman Catholics for whom the Church, local rituals, and processions were until recently an essential part of their social and spiritual well-being.

In the Basque Country linguistic distinctions are made between the house (*etxea*), its inhabitants (*etxenkuak*), and the property (*etxaltea*) to which the house – among other things – belongs. Since the relationship between the house and the *etxenkuak* has been explored elsewhere in depth (Ott, 1989) I will give here only the bare outlines of it.

First, it is essential to appreciate that a Basque house is not simply a physical structure; in Sainte-Engrâce it is also an indivisible social, moral, and spiritual entity that has certain attributes of its own which the *etxenkuak* may use.[6] Among these are the house name, from which the social and spiritual identity of both house and inhabitants is derived, and the house space on the church floor where the female heads of household place their chairs during Mass. The domestic group also have access to the tomb of the house, in which all *etxenkuak* are buried.

In the community people are known and know themselves by the name of their house. In both secular and spiritual matters – such as helping neighbors during the maize harvest or attending a wake – people take part not as individuals but as representatives of their respective houses. The close bond perceived to exist between a house and its members is not, however, confined to the living, but is extended to include the dead; for all those who were members of the house at the time of their death are treated as part of the *etxenkuak*, or domestic group.[7] With the help of prayers and blessed candles the female heads of household are charged with preserving this spiritual bond between house/ inhabitants/souls of the dead/church space/ house tomb.

Traditionally and until recently, the Sainte-Engrâce house was the spiritual and physical center for a number of ritual activities: baptismal, First Communion, and wedding feasts, and a range of mortuary services provided by certain neighbors. Until the 1960s the house was also represented regularly by one of its women in the blessed bread ritual, where women – and by extension their houses – exchanged in a fixed order life-giving prestations of bread (cf. Ott 1980).

As a spiritual, social, and moral entity a house has two vital qualities that inhere in it so long as it has inhabitants: *indarra* – here a sort of spiritual energy or essence – and *zuzena*.[8] Without tangible manifestation itself, the *indarra* of a house is conceived by some as energy, the effects of which no human can actually feel, as one would the shock of an electric current.[9] Whether thought of as spiritual energy or essence, *indarra* connects the house – symbolically – with its tomb, space in church, living members, and souls of its dead. Deemed essential to preserving the spiritual and social unity of the domestic group itself, this *indarra* becomes accessible to the person who inherits the house the moment he or she gains full ownership rights to the property. If the *indarra* of a house is "violated" – in circumstances I will describe later – it is expected that conflict within the domestic group will occur. No house has more *indarra* than any other and the absence of any system of ranking, based on the amount of *indarra* houses are perceived to have or on any other criteria, is wholly consistent with local emphasis on equality (cf. Ott, 1989).[10]

If *indarra* is a spiritual essence or energy that links symbolically house/ inhabitants/church space/tomb and, socially, house/domestic group, what part does *zuzena* play? *Zuzena* and *indarra* may both be used to convey the English terms power, authority, and influence; yet *zuzena* is clearly not the same thing as *indarra*. For a start, *indarra* characteristically supersedes *zuzena*. As one example, a step-parent can have *zuzena* over step-children but not *indarra*, i.e. his or her authority is partial; and it would be thought highly improper if a step-parent behaved as though he or she had *indarra* over them, e.g. by trying to take total control of them. The authority that such control requires can be legitimately claimed and exerted only by a parent. Within the context of the house, *indarra* is thought to have far greater importance than *zuzena* in the maintenance, and the disruption, of amicable social relations. People talk about the *zuzena* of a house as if it were a good influence which, in a manner no one could describe, affects positively the domestic group; but they become confused when pressed for comments about the way(s) *zuzena* makes itself manifest. Unlike the *indarra* of a house, which becomes agitated when conflict occurs in the domestic group, its *zuzena* does not appear to be sensitive to events in the household and is passive in contrast to the energy that is *indarra*. Thus there is perhaps a certain complementarity between the two.

In order to place these two qualities in more concrete, accessible terms, let us turn to a discussion of the people who actually benefit from them – the domestic group – and to the local inheritance practices that bring *indarra* and *zuzena* into play.

In its ideal form the Sainte-Engrâce household consists of husband and wife, their son and his wife, and the unmarried children of that union. Within the context of the domestic group *indarra* informs three relationships: that which exists between the person who inherits the property and his or her siblings; between elder and younger heads of household; and between husband and wife. In all three relationships *indarra* is a quality of relationship that concerns power and equality.

In keeping with Basque notions about sexual equality, both men and women are eligible to inherit; but there is in Sainte-Engrâce a preference for transmitting ownership rights to a son. In accordance with Souletine customary law, inheritance is impartible; but before this ideal is achieved the French rule of partible inheritance must be followed. To satisfy French law, retiring heads of household normally divide their property equally among all their offspring before one of them has been declared the successor. In this contract division of the property is generally fictitious so that "the *indarra* of the house can remain whole" and accessible only to the person who inherited the property.[11] In ideal circumstances, retiring heads of household are succeeded by the child whom they deem to be "of the house" – that is, the person socially, morally, and physically most fit to fulfil the obligations of an active head of household.

For convenience, let us assume that the "child of the house" is a son and that ownership rights are transmitted to him in the ideal manner. When he declares his intention to marry, and provided that his parents are of an age and an inclination to retire, that son can expect his parents and siblings to make a second contract that will satisfy Basque customary law. In this contract his siblings relinquish their claims to the property and in return are guaranteed compensation which, if they accept, is made in cash. In this way the relationship between heir and non-heirs is said to be "equalized," "put in order" (*bardintu*).

As a result of the second contract the heir becomes "the root of the house" (*etxenko zaiña*), of which there can be only one at any given time; and from that moment he is able to partake "of his share in the *indarra* of the house" (*haren etxenko indarren partia*) and to acquire the *zuzena*, "the good influence," of the house. In recent times it has become common for a third contract to be made, soon after the marriage of the heir or heiress, through which the in-marrying spouse gains access to the *indarra* of the house and acquires its *zuzena* –

though the document itself simply states that he/she has become joint owner of the property.[12]

At this point *indarra* starts to cross conceptual boundaries. The inheritor and nowadays his spouse partake of the spiritual *indarra* of the house; but through their pre-marital/inheritance settlement contract(s), and for the in-marrying spouse a dowry, they also acquire jural *indarra*. To acquire legitimately this *indarra* is to gain complete control of the house, the household, and the rest of the property; it is the ultimate power to control the management of domestic resources, the labor of the domestic group, the farm and flock. Once again, *indarra* supersedes *zuzena* as power, authority, and influence; for *zuzena* is not, as *indarra* is, "compelling, persuasive, and irresistible" and, when properly obtained, essential to orderly social relations.

One of the best ways of seeing how *indarra* informs relations of power and equality is to examine cases in which the spiritual *indarra* of the house has been violated. By doing so we will be able to see more clearly how legitimately acquired *indarra* links, symbolically and conceptually, a spiritual essence/energy with the maintenance of order in the secular realm; and, by contrast, how *indarra* illegitimately obtained creates social disorder.

There are three circumstances in which the *indarra* of a house is violated: when the siblings of the heir/heiress refuse to relinquish their claims to the property; when the elder heads of household refuse to retire, even though their successor has married; and, third, when a person other than the heir or heiress is granted legal ownership of a room, or any other space inside the house. In all three circumstances the *indarra* of the house is said to be "in disorder" (*barreatürik*) and is conceived to remain that way until the house and the rest of the property come once more under the control of one person.

At present nine out of ninety-nine houses in Sainte-Engrâce are "in disorder" because siblings (or more distant kin) have failed to cede their claims to the property. Since I have written elsewhere (Ott, n.d.) about conflict between siblings over their inheritance, it is enough to say here that the most intense quarrels and feelings of ill-will in Sainte-Engrâce society spring from attempts by siblings (and sometimes cousins) to control their natal house and property.

The most protracted and bitter dispute in the community has spanned three generations. Towards the end of the last century the heads of the house Behigoitia divided their property equally among their six offspring. For reasons people have forgotten (perhaps deliberately) these siblings were "not in accord" about reuniting the property; but, as is commonly done in cases like this, the son whom the parents had chosen to succeed them married and lived in the house "as if he were its owner." Because the siblings of this man refused to give up their claims to Behigoitia, the initial division and the dispute to which that gave rise were carried to the next generation. The two children of the appointed heir each received from him a "three-quarter share," while five of their first cousins each received one full share in the property.[13]

When they married, three of those first cousins offered to relinquish their claims to end what was becoming an increasingly bitter struggle for control of Behigoitia and which centers on the brother and sister whose father ought to have owned the entire property. The sister had been named heiress of Behigoitia; but when she and her parents violently opposed the marriage of her brother to a local woman, the brother refused, in retaliation, to give his sister his own share in the property. At present these two siblings, now in their sixties and not on speaking terms, and five of their first cousins jointly own Behigoitia; all seven individuals partake of the *indarra* of that house which they have kept in disorder. But not one of them has any jural *indarra*, not even – as her brother once observed with some satisfaction – the woman who lives in the house as though she were its legitimate mistress.

In what way has this struggle for jural *indarra* and ownership of a property concerned relations of equality? For the Basques of Sainte-Engrâce, the relationship between siblings is ideally one of equality. Parents are expected to raise their children as equals and to refrain from favoring the child whom they would like to succeed them. The point at which this is no longer possible is one of greatest tension – when one child is selected to inherit the property. In order to put in order, to equalize (*bardintu*) the relationship between heir/heiress and his/her siblings, parents and their successor must compensate the non-heirs in cash. If, as in the case of Behigoitia over three generations, siblings do not relinquish their claims to a divided property, relations of equality are thrown into disorder; siblings compete for control; and the spiritual *indarra* of the house is violated.

Within the household *indarra* also informs relations of power and equality when the elder heads transfer full ownership rights to their successor. When they do so the shift of power from elder to younger heads is dramatic and abrupt, leaving the elder ones legally *indarrgabe*, "without *indarra*."[14] I was able to observe this in one household, of which I was a member for almost one year. Before her son, the heir, married, the widow enjoyed complete control of the house and farm; but when she transferred full ownership rights over the property to her son, shortly before his marriage, she announced that she was no longer mistress of the house and so no longer had any *indarra* in it. Her son was "master and root of the house"; he alone would partake of the spiritual *indarra* and *zuzena* of the house and would enjoy jural *indarra* as its head.[15] When her son brought his bride into the house the widow made a concerted effort not to behave as if she had *indarra*. Although her first attempts were amusing and unconvincing, she soon mastered her new role as retired female head of household, and managed to teach her young successor how to behave as if she had *indarra*.

During the first few weeks after her marriage the girl rarely spoke, worked hard, and observed closely the relationship between her husband and mother-in-law. One month later the young mistress of the house – at the age of sixteen – was behaving remarkably like her mother-in-law when the elder woman had been in charge. The widow noted with approval the change in her daughter-in-law. "Now she has *indarra*, so you'd better listen to what she says."

From remarks like this it becomes clear that jural *indarra* conferred on a person through a contract will not alone carry much weight in a Basque household. Young heads of household learn how to have *indarra* by learning how to take control of the household and farm. In my own house there was no struggle for *indarra*, no competition for control – though tempers and wills sometimes clashed. As the widow explained, her relationship with the young couple had been "put in order, equalized" (*bardintu*) owing to an exchange: the young couple had received the property and all the *indarra* this conveys, and in return they gave the widow usufructory rights in the house plus the security of being looked after in old age.[16] That the two parties received unequal material rewards was of no importance; for to "equalize" a relationship is to maintain order by investing *indarra* in the socially, jurally appropriate persons.

When an elder head or heads of household do retain practical and legal control of a property after the marriage of their appointed successor, their relationship with the younger couple is "without equality" (*bardin-gabe*) and the spiritual *indarra* of the house is thrown into disorder. When this occurs it is generally owing to the relatively young age of the elder couple, who may be in their forties and not yet ready to be succeeded, though "the child of the house" is married.

In the one case I know well the elder couple of the house Bidabe are in their late fifties and have complete control over all decisions concerning the welfare of household and farm. Their daughter (and only child) has been married for seven years to a local man who works for the electricity company in the lowlands and who lives in Bidabe only at weekends.[17] At thirty-eight, the younger male head of Bidabe is reported, by his wife, to be ready to take control from the elder couple, and has made it known that he would sell the sheep belonging to the property. His wife, however, is still content to have her parents remain in charge for a few more years. Although she is now twenty-eight years old and has two children, the young woman continues to be unusually dependent upon her parents, particularly her mother, for advice. Neither in her marital contract nor in her behavior does she have jural *indarra*. Although her parents are now physically ready to retire (by their own admission), they are not yet psychologically prepared to do so; and their daughter is not yet ready to challenge their *indarra*. In my view, the reluctance of the parents to step down as heads of household is almost certainly linked to their own keen desire to maintain the Bidabe flock. For they, like their contemporaries in the community, cannot conceive of life in Sainte-Engrâce without sheep.

In spite of the expressed desire of the young man to take over Bidabe, it is unlikely that conflict between him and his parents-in-law will occur; for though the elder couple have not yet shown signs of wanting their son-in-law to compete with them for *indarra* I expect they will retire as soon as he does so.

In Sainte-Engrâce the legitimate acquisition of jural *indarra* is regarded as crucial not only to the relationship between elder and younger heads, but also to the marital relationship. Like siblings, husband and wife are thought to be "equal–equal" – morally and socially. If, however, a person inherits a property and acquires his jural *indarra* through contracts executed in the proper manner, but his in-marrying spouse is denied such *indarra* owing to an unpaid dowry, their relationship cannot be "equal–equal." All legitimate power and authority would reside in the inheritor, and the in-marrying spouse would be "like a servant," i.e., contributing a considerable amount of labor to the household in return for food and shelter but without any *indarra*. In all the cases known to me, the spouse whose natal house denied the dowry promised in the pre-marital/inheritance settlement contract was never able to acquire jural *indarra* in the post-marital residence; and although these individuals, now in their eighties, no longer feel anxious about being "without *indarra*," each of them experienced hardship when younger because they lacked that highly prized quality.[18]

In one case a woman was denied a dowry by her family, with whom she had quarrelled. Although the woman's husband was the legitimate "master and root of the house" in which they lived, she herself could not lay claim to being its legitimate mistress, a position to which she was, as his wife and "equal–equal" partner, otherwise entitled. The young couple lived with the husband's parents; and there was in fact some uncertainty surrounding the exercise of power by the two female heads of household. The younger woman was "without *indarra*" because she had not provided a dowry; and her mother-in-law had ceded her own right to be the active female head of household, and consequently had relinquished her jural *indarra* when the property passed to her son shortly before his marriage. Behaving as if they had jural *indarra*, both women competed for control of the female domain, the house. In a society where cooperation and harmony in the domestic group are cherished, such intense competition between two strong-willed women inevitably led to the separation of the two couples, a situation which endured until the elder woman died. From then on, the younger woman enjoyed as much power and authority in the house as her husband, though both partners continue to feel bitter towards the wife's family for having denied her the satisfaction of a legitimate jural *indarra*, and for having kept the spiritual *indarra* of the house in disorder for so many years.

The third and last circumstance in which the "energy" of a house is violated more clearly illustrates the relationship between spiritual and jural *indarra* and shows how illegitimately claimed jural *indarra* disrupts social relations. When an individual other than the heir or heiress gains ownership rights to any space *inside* the house the spiritual *indarra* of the house is thrown into disorder. For although the outsider may never actually live in the house, he will be able to partake of its *indarra*, and this is thought to be highly improper. Ownership of a roof tile (commonly granted to non-heirs elsewhere in the Basque Country), a part of the house garden, or anything else that is external to the house does not make the spiritual *indarra* of the house accessible to the outsider. When he does "get inside" by becoming the owner of a bedroom or part of the kitchen, he usually causes conflict in the domestic group.

I know of three cases in which heads of household have had to share the *indarra* of their house with another person, but I will give here the details of only one.[19] In the house Altçay the heir lived with his wife, their children, and his elder unmarried brother. As the sole and legal owner of the property the heir partook of the spiritual *indarra* of Altçay, which he shared with his wife on account of her pre-marital contract and dowry. As is customarily done when an unmarried sibling resides in a house, the heir gave his bachelor brother usufructory rights. In cases like this, the unmarried sibling often conceives himself to be "like a servant" in the house; for he has no jural *indarra* and thus plays no part in making decisions which concern household and farm. Like a servant, he is obliged to work for the house in return for his room and board.[20] I know of only one case in which a bachelor or spinster has been abused or treated as inferior to the heads of household; but for an unmarried adult, being without jural *indarra* "like a servant" is particularly demeaning.

For the Altçay bachelor, one way of overcoming his humiliation was "to get inside the house" by persuading his brother to grant him legal ownership of his own bedroom as a belated compensation for his having relinquished his claim to their natal house when the heir married. At the time of the heir's succession his elder brother was working in the lowlands and asked for no monetary compensation. Soon after his return to Altçay the elder bachelor pressed the heir to "equalize" their relationship as siblings by giving him the bedroom. When he became the legal owner of that room the bachelor was able to partake of the spiritual *indarra* of the house, which was consequently thrown into disorder – symbolically when the contract granting him ownership rights was drawn up. Social disorder came several years later, after the death of the heir, when his son contested the legitimacy of his unmarried uncle's access to the *indarra* of the house. The son, who had succeeded his father as active male head of Altçay, quarrelled bitterly with his uncle; and the two men took their grievances to the notary in the nearby market town. Owing to the contract in which the bachelor uncle obtained ownership rights to his bedroom, the notary declared that the uncle had a legitimate right to partake of the *indarra* of

Altçay. (Unable to speak Basque and not a Basque himself, the notary argued the case in French and reportedly spoke about the *force* of the house.) The young male head of household conceded defeat but continued to complain to his uncle that he had "got inside the house" only because of his father's soft-heartedness. It was not until the death of the bachelor uncle that the spiritual *indarra* of Altçay became "whole again" and order within the house was restored.

Although people focus their concern and criticism on the spiritual *indarra* of a house when they discuss cases like that of Altçay, the real struggle is informed by *indarra* at another level – where people compete for the power to gain or to retain control, through legitimate or illegitimate means, and behave in a manner that contravenes social expectations. Siblings ought to relinquish their claims to the natal property when "the child of the house" is ready to inherit; and this is the socially and morally appropriate moment for them to claim compensation. An elder couple ought to relinquish control of the property to the benefit of their young successors, even though they may be physically and psychologically unprepared to do so. The person who does inherit should not violate the indivisibility of the property by granting someone ownership of a room; nor should an unmarried, adult member of the domestic group who did not inherit attempt to gain jural *indarra* by "getting inside the house." Where people do not satisfy these expectations there is characteristically competition for power and an inequality – which may be personally felt, expressed socially or materially – that may lead to conflict.

I have tried to show how *indarra* shifts from one level of meaning to another, from the abstract level of spiritual essence or energy that represents the unity of house and domestic group to the level of concrete reality, where the socially, jurally appropriate persons may, or may not, be in control.

In order to appreciate more fully the ideological value of *indarra* we must look now to see what relations it informs in the Roman Catholicism practiced by the Basques of Sainte-Engrâce. In the Mass *indarra* is both procreative and protective: it is the "fruit" of the earth and of the vine, the "patronage" of God, the supernatural "force" that transforms ordinary bread into the body of Christ, the "life force" that God gives to people and which returns to Him when they die.

For many elderly people, the "old religion" with which they were raised had *indarra*; it was particularly persuasive, irresistible, compelling, a fact attested to by the great number of people who attended Mass every Sunday, in the zeal with which people practiced their faith, in the respect they showed for the priest and the Holy Sacraments.

As the ultimate form of efficacious *indarra*, the *indarra* of God is the most persuasive, irresistible, and compelling; through the mediation of the priest it controls human lives and events on earth. Although God formerly used his *indarra* to punish sinners, He now employs it only in a constructive manner, leaving its destructive potential to priests, the men through whom God and laymen are brought into contact.[21]

By virtue of his ordination a priest is able to partake of God's all-powerful *indarra*. But the extent to which he employs it, as well as the uses to which he puts that *indarra* are contingent upon his temperament; for as the Sainte-Engrâce people claim to have observed, priests who are quick-tempered, authoritarian, and nervous are much more likely to use their supernatural *indarra* destructively than are mild-mannered, easy-going priests. However, since even these individuals are capable of putting their *indarra* to destructive ends, all priests are potentially dangerous, to be feared, and to be obeyed.

Although every priest is believed to have supernatural *indarra* humans can recognize only the exterior signs of it, in much the same way as Berbers can see only the external manifestations of a saint's *baraka* (Jamous 1981: 193). Quantitatively, some priests have more supernatural *indarra* than others; there are qualitative differences, too, for some priests have a more efficacious *indarra* than others.

Within living memory the priest who had the most abundant, efficacious *indarra* was Etcheverry, a man whom most people were careful not to provoke. Not an even-tempered priest, Etcheverry was anxious to impress upon his parishioners the spiritual subordination not only of man to God, but also of man to priest. Although many people were fully prepared to accept their spiritual subordination to Etcheverry, a few privately held the view that he was a megalomaniac and expressed disapproval (in the safety of their own houses) of his spiritual domination.

During the Mass the supernatural *indarra* of Etcheverry made itself manifest when he delivered the sermon, which was so compelling that "everyone came to hear it." That *indarra* appears not, however, to have had the proper effect on one man, who burst out laughing in the middle of Mass one Sunday. Having stopped the service, Etcheverry admonished his parishioner. When the man laughed aloud a second time Etcheverry "cursed him in the name of God." As tangible proof of his *indarra* the priest caused the man's mouth to become permanently disfigured.

On another occasion Etcheverry punished a man for having worked on the Sabbath without his permission. When he saw the man harvesting his maize Etcheverry cursed him in the name of God. On Monday torrential rains fell and ruined only the crops of the disobedient parishioner. A few years later another man worked on Sunday and challenged the legitimacy of Etcheverry's use of his *indarra* to destroy things. The following day his two cows went "mad" and plunged from their summer pasture high in the mountains into the gorge of Ehüjarre. Etcheverry was held responsible.

When people recall these incidents they explain how foolish these men were, for they must have known well that a priest can use his supernatural *indarra* against those who cross him; and that, in each case, Etcheverry had a legitimate right to do so since he was expected by God to punish sinners.

Though no one knows of a case, a priest may use his *indarra* to compel evil spirits to do harm by giving a "desiccation mass," which has no benediction, is held outside the church, and causes the victim to suffer great pain, shrivel, and die.[22]

The *indarra* of a priest, derived from that of God, may also be turned to constructive uses. By giving special Masses a priest can make his *indarra* bring under control the natural elements – wind, rain, thunder, snow – when these are severe. In a drought he can compel the elements to make rain. With his *indarra* a priest can also banish mice when they invade the houses and barns in winter; exorcise evil spirits from human bodies; counteract the evil eye; and, through acts of blessing, protect from harm the houses, barns, fields, and livestock of the lay community.

As God's servant on earth, the priest localizes the *indarra* of God in unleavened bread during the transubstantiation of the Host, and in ordinary candles, laurel, water, certain embers, and leavened bread when, by blessing them, he transforms these substances into "blessed things" intended for ritual use by laymen. Both efficacious and protective, the *indarra* of God that inheres in blessed candles will cause severe storms to subside and will assist actively a soul on its journey to Purgatory. When blessed laurel is placed on the house tomb its *indarra* from God will protect all the souls of those who resided in the house at the time of death. In a similar manner the *indarra* that inheres in blessed laurel protects the interior of a house when sprigs of the plant are placed in every room (except the storage space on the ground floor). Formerly, the *indarra* in the burning embers blessed by the priest on Ash Wednesday, and referred to as "blessed fire," also served to protect the interior of a house; for the embers were distributed among female heads of household, who placed them in the hearths of their houses.[23]

In blessed candles, laurel, water, and fire the *indarra* of God is gradually used up and must be renewed annually on a specific day. On Candlemas, for example, all blessed candles are taken to the church by a woman of the house to be re-consecrated by the priest. On Palm Sunday, women gather the sprigs of old, "spent" blessed laurel from the rooms of their houses and burn them in the hearth. During Mass on that day they expose newly cut branches of laurel to the priest's blessing and are thus able to replenish an *indarra* from God for their houses. On Holy Saturday women throw into their gardens any holy water kept in the house and obtain from the priest a newly blessed supply of it that has *indarra*.[24] In the case of blessed fire, the consecrated ember may be consumed physically almost immediately on Ash Wednesday; but its efficacious *indarra* from God will continue to inhere in the hearth for twelve months.

Of all the "blessed things" only blessed bread has a God-given efficacious *indarra* that is never exhausted and which is life-giving. Since I have discussed these properties elsewhere (Ott 1980) I will point out here only that the blessed bread of Easter is thought to have more supernatural *indarra* than bread blessed on an ordinary Sunday "because it can find and resurrect the body of the drowned." Proof of this extraordinary *indarra* is sought in accounts of accidental death by drowning, in which the blessed bread of Easter did locate the corpse and physically reunite the deceased with his survivors and house.

In addition to his supernatural *indarra*, transmitted from God to things and people on earth through him, the priest in Sainte-Engrâce also has secular *indarra*, a quality ascribed to him by his parishioners – not automatically, but when he demonstrates possession of it in fulfilling his duties as town hall secretary or in his role as mediator in disputes.[25] Some Sainte-Engrâce priests have had more secular *indarra* than others; and they have earned it in different ways. Father Etxaun, for example, was never particularly interested in his job as town hall secretary. But, during his term of office (1952–8) he was especially successful in settling quarrels between siblings over their inheritance and between young people and their parents when a proposed marriage was regarded by the parents as "too close" consanguineally. A quiet, gentle man, Etxaun had an *indarra* that was persuasive and compelling for those close kin who wanted to be reconciled, but who found they could not do so on their own.

Since he refused to talk to me about his involvement in dispute settlements, the present priest gave me little indication of his secular *indarra* in that sphere; and although people made it clear that "Kiki" (as they refer to him in private) does have *indarra* as a mediator in quarrels, his secular *indarra* is more irresistible, his control more complete in another secular role – that of town hall secretary.

Although the mayor and municipal councillors of Sainte-Engrâce have political power in local government, the priest is acknowledged to have more political *indarra* than any politician in the commune. In Sainte-Engrâce the priest potentially has the power with which to control the relationship between the people and an outside world that includes politicians at cantonal and provincial levels, entrepreneurs in business, tourists, and government bureaucrats; for the role of town hall secretary has come to include public representation of individual and community interests to outsiders. In addition to keeping the cadastral, birth, death, and marriage records, the priest is called upon regularly to assist his parishioners with applications for pensions, child allowances, unemployment benefits, and all other matters requiring fluency in French, familiarity with regulations and with bureaucrats, and the ability to argue a case forcefully – all of which "Kiki" possesses and which make manifest his considerable secular *indarra*.

Although no one challenges the priest's *indarra* when it obtains for laymen what they want from the government, a few men have contested openly its use in promoting community projects which the priest favors in spite of local opposition, such as the construction of a road to Spain and of facilities for tourists. A priest may use legitimately his supernatural *indarra* to destroy the property of a sinner; but, these few unusually outspoken men ask, can the priest legitimately use his secular *indarra* to "develop" the community in a manner that may lead not only to the destruction of property, but also to the destruction of a way of life the majority of people claim to cherish. The majority of the people are, however, unwilling to make public their views about the road and tourists because they "do not like to go against the priest." Their extreme reluctance to challenge the priest, in either secular or spiritual matters, has been made apparent to me on numerous occasions. For example, when the priest abolished the ritual of blessed fire on Ash Wednesday "because it takes too much time" the people were outraged. Rather than build a fire outside the church (from which embers were traditionally taken, blessed and distributed by the priest) the priest instructed the acolytes to pass round white candles, which we held while he blessed them from the altar. The anger of his parishioners intensified when, at the end of Mass, he told us to return the candles to him. In spite of their extreme annoyance about the laziness of their priest and his abolition of what was, to his parishioners, a "lovely custom," the people never expressed their feelings to "Kiki." The complete acquiescence of the people to their priest was also made abundantly clear in the 1960s, when they refused even to ask him why he had made obsolete the ritual giving of blessed bread – a ritual of special significance to them and the loss of which they still lament. As people explain, it would have been morally and spiritually wrong to contest the priest's decision to stop the ritual; for it would have been tantamount to challenging the *indarra* of "Kiki."

Among the elderly parishioners, a fear of the priest's supernatural, potentially dangerous *indarra* often underlies their passive submission to him. For young and old alike, "crossing a priest will bring disorder" – that state of being *barreatürik* – because it upsets the relation of spiritual and secular subordination that ought to obtain between a priest and laymen. In answer to queries about their refusal to challenge the *indarra* of "Kiki," people simply say, "as priest he is *nausi* [master] and to him we are like servants, without *indarra*."

As sorcerer and as God's servant, the priest enforces the spiritual subordination of the lay community to himself through his supernatural *indarra*. With a curse or a blessing he has the power to transform things – the mouth of an irreverent parishioner, a loaf of leavened bread. By virtue of his supernatural *indarra* the priest can control the natural elements as effectively as he can control laymen who contest his role as spiritual "master" on earth.

As town hall secretary the present priest exercises a control in the temporal affairs of men which evinces their secular subordination to him. That control is not total, but its application has important economic and social consequences (in particular, a steady flow of cash into the community and improved living conditions), which in turn serve to reinforce the priest's perceived superiority in the secular realm.

Unlike those relationships in which harmony is possible only if a basic premise of equality is upheld – between heir/heiress and non-heirs, elder and younger heads of household, husband and wife – the relationship between priest and laymen must be one of inequality if order is to be maintained in both spiritual and secular realms; for, by virtue of his ordination, the priest links heaven and earth. In Sainte-Engrâce, the idiom through which that inequality is expressed is *indarra*.

In my analysis of relationships informed by *indarra* a recurring theme has been the close association between the correct use of *indarra*, its investment in the spiritually, socially, jurally appropriate people, and the maintenance of order. Like *baraka*, *indarra* is not, however, always identical with the maintenance of order; for sometimes it transcends and breaks the boundaries of order.[26] In conclusion, I would like briefly to discuss three spheres in which *indarra* is the idiom through which people discuss forms of socialized (legitimate) behavior and forms of unsocialized (illegitimate) behavior; in the first *indarra* is controlled, in the second it is not. The spheres are sexuality, drinking, and a type of dispute settlement called *xikito* (from a Basque term for castration).

At sexual maturity men and women acquire an *indarra* that enables a man "to curdle" a woman, and a woman to conceive (cf. Ott 1979). Some people have more procreative *indarra* than others; and those who show an abundance of this vital quality through the number of their offspring occasion considerable comment in the community.

The man whom most people reckon to have the greatest amount of procreative *indarra* is Iribarne. Having fathered thirteen children he has made manifest his *indarra* and has used it in a controlled, legitimate way for the socially acceptable purpose of begetting children within marriage. Iribarne's *indarra* is socialized, safe, and has produced what it was meant to produce: sons and daughters for the house. While women pity his wife for her thirteen pregnancies and claim she has more procreative *indarra* than Iribarne himself, men openly admire the virility of the man, though most people feel that he might have controlled his sexuality a little more carefully by practicing coitus interruptus for longer periods between his wife's pregnancies.

In contrast to Iribarne there is another local man, Behoçaray, whose procreative *indarra* was, in his youth, reckoned to have been used in an uncontrolled, unsocialized, dangerous manner. When Behoçaray was twenty-one years old he made pregnant a sixteen-year-old girl. Although he wanted to marry the girl, her parents forbade him to do so and sent their daughter to the house of a relative in the lowlands. Behoçaray was judged by the parents to "have too much *indarra*" and to be quite incapable of controlling it, so much so that "he would make any woman pregnant every nine months." When Behoçaray made pregnant a spinster of forty-five a few years later, the parents of the young girl reportedly felt that their charges against Behoçaray had been vindicated. There was no talk of marriage between the spinster and Behoçaray, before or after the pregnancy. A scrawny, unshaven man, still a bachelor, Behoçaray was the butt of frequent jokes during the early part of my fieldwork, when men would caution me against getting too close to him lest his dangerous procreative *indarra* should give me a daughter. (His two bastard offspring were girls, which led people to speculate that his *indarra* was capable only of producing females. In the same vein I have heard a woman tease her husband about having an *indarra* that produces only males, for they have four sons.)

Having ceased to be able to "curdle" a woman Behoçaray is no longer thought to be sexually dangerous;[27] but a total lack of control and unsocialized behavior in another sphere continue to set him apart from the rest of society, for he has become an alcoholic. Though sympathetic neighbors (who are also his close kin) try to introduce some semblance of order in Behoçaray's life – by giving him hot meals once a week, washing and mending his clothing, and looking after his livestock when he is on a "binge" – Behoçaray and his house are perceived to be in a state of perpetual disorder, physically and spiritually; for, as everyone agrees, Behoçaray cannot regain control of himself and his house unless he gives up wine.

Although the Sainte-Engrâce people enjoy drinking several glasses of wine each day, they dislike wine that has "lots of *indarra*," i.e. one that has, through the process of fermentation, an alcoholic content greater than 11 per cent. Normally their wine has only 10$^1/_2$ per cent alcohol, and this people tend to mix with water.[28] Both sexes also enjoy spirits, especially Ricard. For these Basques drinking is a social activity that gives much pleasure at table, but which requires moderation and restraint. Controlled use of the *indarra* in wine and spirits is socially acceptable; its abuse is not.

One last context in which *indarra* is concerned with control and socialized behavior is provided by a form of ritual verbal duelling (*xikito*) among shepherds that is now obsolete. Unlike many of their European counterparts, Sainte-Engrâce shepherds get along remarkably well with each other; they do not steal each other's sheep or accuse each other of trespassing on their summer pasturage. Until recently, *xikito* was an institutionalized means of ensuring that conflict did not occur in the mountains, where shepherds and sheep live in summer. If a man became angry with a shepherd from another syndicate, he was obliged to play *xikito*, in which opponents stood at a considerable distance from each other and shouted rhymed insults – mostly concerned with forms of abuse to the opponent's genitalia or arse.[29]

The object of the game was to exhaust one's own anger and that of one's opponent by composing verses so obscene and absurd as to cause laughter. If a shepherd played *xikito* successfully (indeed I know of no man who did not) he subjected to control the *indarra* of his anger – easily translatable into physical violence – through the rhythm and obscenity of his verses. Restraint of *indarra* may not in this context link heaven and earth; but it too is aimed at control and the maintenance of order.

For social anthropologists of southern Europe, the interest of *indarra* lies not only in its unusual tendency to combine physical and secular realms with spiritual ones; for when we look outside the Basque Country to the Mediterranean, we find there are concepts by no means equivalent to *indarra*, but which appear to be of the same order as this Basque notion. In particular, we find striking affinities between *indarra* and *baraka* – among other things – "the source of all increase and prosperity . . . a kind of energy inherent in certain people," e.g., saints (Maher 1974: 100). In the same manner as *indarra*, some things *are baraka*; others *have* it (Jamous 1981: 202). There are, of course, many equally striking differences between these two concepts. As Ernest Gellner has observed (personal communication) *indarra* is much more pervasive and universal than *baraka*; and, unlike *baraka*, which contrasts with the maintenance of order, *indarra* is identical with it in some spheres (in which *indarra* is legitimately obtained and employed) but transcends it in others (in which *indarra* is illegitimately acquired or wrongly used and, as a result, causes disorder).[30]

Looking outside the Mediterranean, *indarra* also seems to be a concept of the same order as *mana*, though here again there are as many marked differences as similarities. Like *indarra*, *mana* has a range of seemingly disparate meanings and brings together "semantic spheres that are usually opposed – heaven and earth, spirits and men, sacred and mundane power" (Pitt-Rivers 1974: 3).

Rather than treat *indarra* as if it has a *valeur symbolique zéro* – as some have treated *baraka* and *mana* in an approach which cannot do justice to the richness of these notions (cf. Jamous 1981: 204) – I have looked for the significance of *indarra* in the "association it makes between different realms of meaning" (Pitt-Rivers 1974: 6). Although it is outside the scope of this essay to trace any family resemblances between *indarra* and other similar concepts, I have tried to provide Mediterraneanists with an ethnographic and conceptual base against which to contrast, in particular, *baraka* and grace.

Notes

1 My field work in Sainte-Engrâce was conducted over a twelve-month period in 1976–7 and in 1979. Since 1981 I have returned to the commune annually. Although my interpretation of *indarra* is based mainly on material collected in Sainte-Engrâce, it has been influenced by my field work in the Spanish Basque community of Aramaiona (Alava). Earlier versions of this paper were presented at seminars for the social anthropology departments at Oxford University, the Queen's University Belfast, and Manchester University. I wish to record here my gratitude to those audiences for their helpful comments and questions; but my greatest debt is to Professors Ernest Gellner, the late Emrys Peters, and especially to Julian Pitt-Rivers for having stimulated my thinking in new directions.

2 An alternative spelling is *indar*. In some texts, e.g. the Bible, *indar* and *indarr*, as radicals, are used interchangeably.

3 Barandiarán (1979: 67) reports that in some parts of the Basque Country the souls of former members of a house are thought to be violent gusts of wind, which he glosses as *indar*.

4 See in particular Gellner (1969, 1981), Jamous (1981), Maher (1974), S. M. Peters (1952), and Westermarck (1926: 47–9, 80, 124–8), especially on the *baraka* of certain mountains, the moon, mad men, and on the various ways of increasing the *baraka* of one's harvest in Morocco.

5 Jamous (1981: 203) reports that it would be difficult to talk at great length or in abstract terms about *baraka* to the Berbers of the Rif. The two concepts about which Basques talk more easily than *indarra* are *üngürü* (rotation) and *aldikatzia* (alternation, serial replacement). Cf. Ott 1981.

6 This is a common characteristic of the Pyrenean house (cf. Leroy Ladurie 1978: 59–60).

7 A close, immutable bond between living and dead members of a house is also reported by Leroy Ladurie (1975: 61) for Montaillou and by Bourdieu (1972a) for the Kabyle.

8 Interesting parallels are afforded by the Montaillou house, which has its own *astre*, *sa bonne fortune* in which both living and dead members participate (Leroy Ladurie 1975: 60), and the Kabyle house, which has *baraka* (Bourdieu 1972a). Both authors report great concern among the survivors that the deceased should not take away from the house its *astre*, *bonne fortune* or *baraka*.

9 The person who suggested this to me was surprised to learn that *indarra* denotes electricity in some parts of the Basque Country.

10 The equality perceived to exist between houses contrasts sharply with reports from the neighboring province of Béarn, where houses are ranked according to – among other things – material wealth (cf. Bourdieu 1962, 1972b; Rieu-Gout and Sauzéon-Broueilh 1981).

11 Although I heard no one else quantify *indarra*, one woman thinks that parents retain one-quarter of the *indarra* of the house and convey to their children the remaining three-quarters.

12 Pre-marital inheritance settlement contracts are examined in greater detail in Ott 1989: 252–4.

13 The appointed heir is reckoned to have received one and a half shares in the property from their parents, whereas his siblings got only one share each. The heir received the extra "half share" because he was "the child of the house."

14 Traditionally the elder in-marrying spouse was in a precarious position if the

inheritor died first; for he/she was thought to have no right to remain in the house. See Ott 1989: 261–2.

15 At least until his wife persuaded him that she ought to be granted jural *indarra* in yet another contract. Their pre-marital/inheritance settlement contract did not give her *indarra* because she did not provide a dowry.

16 The ease and graciousness with which the widow accepted her new role may have stemmed in part from the fact that she receives a substantial pension, upon which the younger couple depend and which gives the widow a certain amount of power.

17 The arrangement whereby a father or father-in-law acts as shepherd/farmer while his son or son-in-law earns money in the lowlands is becoming more common.

18 Even as jural *indarra* is acquired through a contract, a dowry (in the case of the in-marrying spouse), and correct behavior, there is one circumstance in which it can be lost owing to incorrect behavior – namely, if a widowed in-marrying spouse remarries. This is the reverse of the traditional view that such a person lost *indarra* at the death of his/her spouse and could only regain it by marrying a sibling-in-law or more distant affine who had a legitimate claim to the property. See Ott 1989: 261–6.

19 Details of the other two are provided in Ott, 1989.

20 Until the 1950s, it was customary for Sainte-Engrâce children to work as *domestiques* from the age of twelve until they were about sixteen years old. Most children found employment in local houses, in which they were given a bed, meals and a small annual stipend.

21 As Ernest Gellner (personal communication) has observed, the reverse is true in contemporary Islam. Formerly God worked through his agents, whom He endowed with *baraka*, whereas nowadays He works without mediators.

22 To the horror of others present, one old woman once advised me that a local man had in fact asked the priest (a predecessor of Etcheverry) to give a "desiccation mass," but the priest refused.

23 For a summary of contexts in which the *indarra* of blessed things is protective and efficacious, see Ott 1981.

24 Some elderly people say that holy water ought to be thrown into the garden so that any trace of *indarra* which has not been used up can increase the fertility of the soil, encourage the growth of vegetables, and return to the people of the house in their food.

25 In the French Basque Country it is highly unusual for the priest to hold the office of town hall secretary. The Sainte-Engrâce people do not seem to know that it is they who confer secular *indarra* on their priests; they do know that it is not God-given.

26 I am grateful to Ernest Gellner for having suggested this feature of *baraka* to me.

27 Both sexes are thought to experience menopause (cf. Ott 1979).

28 The priest and I were among the few who enjoy wine with "lots of *indarra*" and were left to drink the vintage wines at baptismal, First Communion, and wedding feasts which other guests found "too strong."

29 The *xikito* "verse" ought to be four-lined, but Sainte-Engrâce shepherds did not always adhere to this rule. Below is an example of *xikito*, recorded by Irigaray (1963: 5).

Xikito dioka xikito,
ehiz hortan kito;
gure artzan horak

ehün khana pito;
balü ere ehun haboro
bat aski dik beretako
bestiak oro hiri üzkütik sartzeko. Xikito!

Xikito, you say, xikito,
you are not yet free;
our sheep dog over there
has a hundred fathoms of penis;
he would have it another hundred more
one is enough for him
all the rest for shoving up your arse. Xikito!

30 Jamous (1981: 204) notes that honor is "aussi force abstraite et concrète, qui fonde un ordre et le transgresse."

References

Azkue, Resurreccíon María de 1935. *Euskaleŕiaren Yakintza*, Vol. I, Madrid, Espasa-Calpe

Barandiarán, José M. de 1979. *Mitología Vasca* (3rd edition), San Sebastian: Editorial Txertoa

Bourdieu, Pierre 1962. Célibat et condition paysanne, *Études Rurales* 5–6 (April–September), 32–135

Bourdieu, Pierre 1972a. *Esquisse d'une théorie de la pratique*, Geneva, Librairie Droz

Bourdieu, Pierre 1972b. Les Stratégies matrimoniales dans le système de réproduction, *Annales*, 27 (4–6), 1105–25

Gellner, Ernest 1969. *Saints of the Atlas*, London, Weidenfeld and Nicolson

Gellner, Ernest 1981. *Muslim Society*, Cambridge, Cambridge University Press

Irigaray, Léopold 1963. Chiquitoak, *Igela: Euskaldun Heterodozoen Errebista* (March–May), 5–8, Paris, Rédaction Direction

Jamous, Raymond 1981. *Honneur et "baraka" : les structures sociales traditionnelles dans le Rif.* Cambridge, Cambridge University Press; Paris, Maison des Sciences de l'Homme

Leroy Ladurie, Emmanuel 1975. *Montaillou: village occitan*, Paris, Gallimard

Maher, Vanessa 1974. *Women and Property in Morocco*, Cambridge, Cambridge University Press

Ott, Sandra 1979. Aristotle among the Basques: the "cheese analogy" of conception, *Man*, 14, 699–711

Ott, Sandra 1980. Blessed bread, "first neighbours" and asymmetric exchange in the Basque Country, *Archives Européennes de Sociologie*, 21, 40–58

Ott, Sandra 1981. *The Circle of Mountains: A Basque Shepherding Community*, Oxford, Clarendon Press

Ott, Sandra 1989. Mariage et remariage dans une communauté montagnarde du pays basque, in J. G. Peristiany (ed.) *Le Prix de l'alliance en Mediterranée*, Paris, Editions CNRS, pp. 249–70

Ott, Sandra, n.d. Equality and equilibrium in the Basque Household.

Peters, Stella M. 1952. A Study of the Bedouin (Cyrenaican) Bait, unpublished B.Litt. thesis, University of Oxford

Pitt-Rivers, Julian 1974. *Mana*. Inaugural Lecture, The London School of Economics, Welwyn Garden City, The Broadwater Press
Rieu-Gout, A.-M. and Sauzéon, M.-L. 1981. Parenté et alliance dans la Vallée de Barèges. *Ethnologie Française*, 11 (4), 343–58
Westermarck, Edward 1926. *Ritual and Belief in Morocco*, Vol. I, London, Macmillan

11

Postscript: the place of grace in anthropology

JULIAN PITT-RIVERS

Prologue

In the introduction to this book we observed that the word "honor" had entered anthropology only in the 1960s and we gave some considerations to account for this, to my mind curious, lacuna. The word "grace" is today in a condition somewhat comparable to "honor" in the 1960s and it is one of the aims of this essay to endow it with the recognition it deserves.

I have found only three authors who have called attention to its existence as a general anthropological concept – briefly only[1] – and none who has undertaken a detailed examination of its logic and potential usage.

This, despite the enormous importance of a concept of this order in Christianity, Judaism and Islam, the huge theological literature on the subject – (surely the anthropology of religion can no more ignore Western theology than the anthropology of law can ignore Western jurisprudence?) – and the fact that a large number of Christian heresies were provoked by disagreements as to the nature of grace, from the Pelagians onwards. But blindness in this matter can hardly be attributed simply to the parochialism of modern university disciplines or the social scientists' distrust of the expertise of the theologians: no anthropologist has to my knowledge asked himself whether there is anything remotely equivalent to grace among the concepts of Buddhism, Hinduism, Shintoism, or Taoism, though the sinologists have used it freely to translate the extensions of the word *fu* (in origin a sacrificial offering, but also, the return of grace, thanks), whether the problem to which the doctrines of grace tender an answer has no echo beyond the religions of the Book, let alone the peoples without writing, or whether grace can be treated as a universal concept or only as an element of Western culture, a question which I asked with regard to "honor" in the introduction to *The Fate of Shechem* (Pitt-Rivers 1977).

Personal destiny, with which grace is very much concerned, looms large in the works of Evans-Pritchard, both on Zande witchcraft and on Nuer religion, yet he who favored employing Catholic theological terms as the nearest equivalents to the concepts of the Azande or the Nuer, seems never to have used the word.[2] But these are all petty reasons for astonishment in comparison with the immense importance of the derivatives of grace outside the realms of theology, in the notion of gratuity. (At least gratuity is an abstract, universal, and theoretical concept.)

Not only is "grace" said before and sometimes after every meal in an Oxford college, I doubt whether any Englishman can get through a single day of his life without saying "thank you" at least a hundred times. According to the Oxford English Dictionary, "Thank" is: "1) Thought (M.E.) [It comes from the same Old Teutonic root as "think"]. 2) Goodwill, graciousness, favor. 3) Grateful thought, gratitude. 4) Expression of the grateful acknowledgement of a benefit or favour."

Is not "thank you" in the "hundred-times-a-day" sense, simply the recognition and acceptance of a gratuitous gesture? The romance languages connect the notion rather more explicitly with grace: French, *merci*; Spanish, *gracias*; Italian, *grazie*, etc. It may be that this notion is not evoked when saying thanks because the Old Teutonic root (thank), used in casual conversation is not recognized as having anything to do with it. The romance root (*gratus*), which recalls the word "grace," is reserved for more formal occasions when gratitude may really be felt. Yet it remains an enigma that the notion of grace should have escaped the anthropologists for so long. This is only the more remarkable in view of the attention they have given, in recent decades, to the problem of reciprocity. Can one explain systems of reciprocity adequately without considering the possibility of non-reciprocity, i.e. gratuity? Reciprocity is the basis of all sociation, in the form of systems of exchange, of women and of food, of labor and services, of hospitality and of violence. Anthropologists attach importance, rightly, to the detail of personal conduct in their understanding of human relations. And are not thanks the common coinage of encounters between persons? Yet what do they imply? What is their logic? Not even Goffman nor the ethnomethodologists have told us.

This oversight might be accounted for simply by a regrettable tendency among those of a functionalist turn to mind to jump to conclusions regarding the significance of human actions on the basis of expressed intentions, without examining their mode of expression; to reduce each institution to "what it amounts to" or "what it does" in terms of practical results, ignoring its cultural roots, thinking that there is nothing more to be known about the culture of a people than what they themselves consciously recognize. Such accusations cannot be leveled at more than a minority of the profession today. That grace should have been ignored for so long is still puzzling.

It is true one does not think of theology every time one says "thank you," or wonder why the word "grace" and its derivatives have such diverse connotations in different settings: why grace is engraved on English coins (Elizabeth II REG. D.G.) as well as said before a meal, why it is used in writing to a duke as well as bestowed on the congregation at the end of Matins, why it is invoked to applaud a dancer as well as to explain the success of a politician, why it names the three pagan goddesses concerned with personal endowment as well as voluntary embellishments added to a musical score, why it is left on the café table for the waitress as well as implored of God in the hope of salvation. One is justified in wondering whether it is still the same word or has become through the wear and tear of time a virtual homonym (Pitt-Rivers 1967). Yet this semantic richness does not mean that there is no association between the different senses.

The meanings which a single word has in different contexts, or had in the forgotten past, are guides to the premises which underlie its daily conscious usage, but daily usage is indifferent to contradictions arising between its various senses, and leaves them to be sorted out at the level of action. (This is the case of honor also.) Thus it is not necessary to analyze a word in order to know how to use it correctly.

Moreover, the implications of the concept differ from place to place; expressions of gratitude are wont to follow different rules of etiquette in different social milieux, as in different cultures, and it is only by reflecting on all potential differences that one comes to understand the concept as a whole, and I maintain that grace *is* a whole. Thanks can in some situations be interpreted as a reluctance to pay what the recipient expects as his due; in others, on the contrary, to thank is to recognize indebtedness and represents a promise to return the gift or service in the future. (Such cultures might be described as "Maussian," for they have already understood "the necessity to return presents," as Mauss put it in the subtitle of his great essay on the gift.) But the etiquette concerning the return of gifts contains more subtleties than Mauss explained.

The only general rule that can be cited is that grace is always something extra, over and above "what counts," what is obligatory or predictable; it belongs on the register of the extraordinary (hence its association with the sacred). Nevertheless whenever a favor has been done the return of grace is always expected, whether in the form of a material manifestation (regardless of the material value of that which is returned) or merely in verbal expression.

However, there are also conventional formulae for the acknowledgement of thanks which follow the same principle in all European languages. They consist curiously enough in denying that a favor has been done. Hence in Spanish *de nada*, in Italian *de nulla* and in French *de rien* (though this is considered undistinguished; it is more elegant to say *c'était un plaisir* or *je vous en prie* as in Italian *prego*, in German *Bitte*). In English "Don't mention it," but in American English "It was a pleasure" is the most explicit form. The denial is not in fact the denial of the sentiment which inspired the act of grace but rather that any obligation has been incurred. It is a way of asserting that the grace was real, that the favor was indeed gratuitous. It is a guarantee of the purity of the motives of the gratifier, a way of saying (as we shall see below) you *owe* nothing for this favor, it is an act of grace. Without wishing to sound cynical, I must nevertheless point out that by, so to speak, stressing the gratuitous nature of the gesture by denying that any obligation has been incurred, it not only maintains the purity of the motives of the gratifier, it maintains his moral supremacy, which is not to be modified simply by verbal thanks but leaves him a creditor, should the occasion ever arise where a more serious return of grace is possible.

The modern tendency in England to say "thank you" when no thanks are conceivably due, almost as a pause-word when one does not know how otherwise to bring the conversation to a close, appears to me to impoverish the notion and things have gone so far that it has now become common custom in commercial aeroplanes to use the phrase to terminate an emission over the intercom system, indicating thereby that the unknown, unseen, uninvited speaker has finished what he had it in mind to announce and is switching off. He thinks perhaps that he is giving thanks for the attention that has been paid to his message. But little does he know it when there is an anthropologist in the cabin who responds to this proffered hypocritical gesture of grace by cursing the interruption of his reading or his snooze. (A curse is, of course, an expression of negative grace – like witchcraft as we shall later see – and on that account alone is worthy of attention.)

Such excessive use of the catch-phrase of thanks does not always ingratiate the thanker and among the simple folk of southern Europe to thank when no thanks are due was commonly regarded as "uncouth" behavior and raised suspicions as to the speaker's ability to recognize when gratitude *was* due.[3]

This devaluation of thanks, encountered in British or American airliners is, no doubt, inspired by the feeling that some attempt at sociation is due between the crew and the passengers in their charge, even if one cannot determine what: a wave of the hand or a wink suffice in some contexts to establish it, but the trouble with intercom systems is that you cannot actually see your interlocutor. Reciprocity is the essence of sociation, as has been said, whether in mechanical or organic solidarity, whether in exchanging salutations or in business and whether the return be immediate or protracted; it is, as it were, the cement that holds any society together, for it establishes relations between persons; once you have exchanged something, you are related. The often meaningless salutations on the path have no other function than this (Firth 1973).

Exchange is not merely an economic fact making the division of labor possible and enabling peoples of differing ecology and cultural achievement to make their products available to each other and thus extend the potentialities of their lifestyle, it is above all a social fact, for, as a principle, it governs much more than the exchange of "useful goods" and functions: one can exchange pleasures, woes, secrets, women, insults, vengeance, hospitality, conversation, stories or songs, and above all gifts, but the economic justification for exchange is, as often as not, subsidiary or lacking altogether, even in the exchange of goods, as in those Somali marriages, studied by I. M. Lewis (1961: 139), in which camels, paid as bride-price, return as part of the bride's dowry to their original owners. Yet morally they are no longer the same camels. For something of the giver always accompanies the object exchanged, and if in the first place the camels were exchanged for the bride, in the second they testify to the rights of her kinsmen in regard to her children.

That the principle of exchange reigns over many more aspects of society than the economic has been recognized ever since Mauss' essay on the gift, and Lévi-Strauss later applied the principle to kinship and marriage, stressing the importance of the groups between whom exchange takes place. This was the point of departure for the structural study of kinship. Yet an aspect that has attracted less attention is that which determines the ways in which, and the extent to which, reciprocity is evaluated and equivalence is established. In the elementary structures of kinship there is no problem: one woman is fair exchange for another. But in other forms of exchange the degree of specificity varies; one man's labor is known in advance not to be as productive as another's, even though this can rarely be admitted openly, yet when it is a matter of exchanging goods their relative values are discussed freely. This is the basis of all bargaining and indeed of commerce itself. But this is not all: the mythology of our times assumes that agreement as to economic equivalence is the condition of all exchange, and to establish this, some kind of estimation must be made. Yet, in fact, such equivalence is necessary only if the aim is commercial, not if it is moral, for example in the field of religion, as the parable of the widow's mite made quite explicit.

There are also reciprocal relations which do not involve a contract of any sort, nor any estimation of value, for they are between participants whose interests are solidary (in the legal sense that all responsibilities are shared and all benefits are held in common). Nothing is specified by way of a return. The nature of the relationship suffices to determine what is due and when. This is what Meyer Fortes called "kinship amity," "where there is no mine nor thine but only ours." The basic distinction between the parties involved, which is essential to the notion of commerce, is missing and in the field of reciprocity we must therefore differentiate between relations of equivalent exchange and relations of amity, between contract and benevolence, between precise commitments, entailing estimations of equivalence, and mutual aid whensoever needed.

Yet beyond the realms of kinship amity where all is held in common and without entering the field of exchange as the law defines it ("a mutual grant of equal interests, the one in consideration of the other," Blackstone in O.E.D.), it is possible to speak of the exchange of favors. Both favors and contracts involve reciprocity, but contractual reciprocity, the basis of trade, is not the same thing as the reciprocity of the heart. The latter escapes the attention of economists (and fiscal authorities, though they have introduced "gift tax" in the attempt to frustrate the evasion of other more onerous taxes). The ties of kinship or of friendship, the hidden incentive, the act of forbearance, the incalculable benefits of favorable decisions, the aim to please, evade the notice of both, for they cannot be evaluated, yet at the micro-level that concerns the anthropologist they are primordial, the very tissue of social relations, and in the end they bear fruit at the macro-level of society and distort the calculations of those who suppose they can be ignored.

Hence, for example, the problem of "corruption" in Latin America has received much comment, especially from foreign analysts who have tended to deplore it as a vice in the administration, if not in "the Latin-American character." The *mordida*, literally "the bite" (and the word is used in syndical affairs in the United States in quite the same sense) is sometimes blamed for the various ills of the Mexican economy; it is indeed quite current that those empowered by the State to give their assent to the operations which are subject to legislative control expect to be thanked for giving it, for personal relations remain personal in Mexico regardless of the legal function of the participants. This "customary expectation of the return of favors" may be reprehensible from the viewpoint of modern Anglo-Saxon values, yet what could be more natural than the principle from which the *mordida* springs: that a favor requires to be returned, and all the more imperatively when the possibility is open that it may be withdrawn if reciprocity is not forthcoming. In Mexico, as in most of the non-European world, the values of the heart are rated higher than those of the law, and the confrontation between the two leads sometimes to anomalies, as in the case of that Governor of the State of Chiapas who ruined his political career by refusing to accept *mordidas*. Apart from the moral pretensions of his stance which were resented by those he governed, it was felt that such a man could not be trusted in personal relations. Moreover, is it not also frequently the case in Europe that the values of the heart take precedence over legal and economic considerations? The example of neighborly cooperation between small peasant farmers will be examined below. Indeed, wherever the idea of gratuity appears, contracts, calculation, and legal obligations step into the back seat and the counter-principle of grace takes over. Yet the back seat is never empty and from the moment that grace reigns over the order of the day, the clever ones start calculating the odds and exploiting the situation to their advantage. Hence it comes as no surprise to learn that there are plenty of people in Mexico who strongly resent having to give a *mordida* to an official who is already paid to do his job, yet if they are not quixotic they give it all the same. But since the servants of the state are poorly paid in Mexico it is difficult to see what advantage anyone would have in entering its service if he could not look forward to making something on the side. Thus, as things stand at present, those who require the State's services are those who pay for them. What could appear more equitable? However, applied at a higher level, of government, rather than of petty local administration, these principles produce a rather different result: the concentration of power in the hands of a political class whose existence is not recognized by the ideology of the State. Thus, behind the façade of administrative law the personal policies of the administrators reign and the values of the heart are exploited by their calculations of how to increase their fortune or their power.

There are, then, two parallel modes of conduct, ideally, even if they are not always easily distinguished, which correspond to the old opposition between the heart and the head: that which is felt and that which is known, the subjective and the objective vision of the world, the mysterious and the rational, the sacred and the profane. They are governed, respectively, by the principle of grace and by the principle of law, that is to say, predictable regularity, as well as justice and the law which impose order in human affairs – from which pardon (grace) authorizes a departure. Under the heading of "grace" it is possible to group all the phenomena that evade the conscious reasoned control of conduct.

Etymology: theological origins and extensions

Let us look then first of all at the notion of grace in general before dealing with its different aspects. From the Greek root "Charis" (χάρισ), we get charity, charisma, eucharist and so forth; from the Latin, derived from it, gracious, gratitude, congratulations, etc. The origin of the word is religious; it is a theological idea which has found various spheres of extension outside the realm of theology. Benveniste (1969: 199) traces it to an Indo-Iranian root: the Sanskrit *gir*, a song or hymn of praise or of grace, or to give thanks. He notes also that, like all other economic notions, its economic sense derives from the totality of human relations or relations with gods rather than the contrary. Gratuity is the core of the notion, that which is undertaken not in order to obtain a return but to give pleasure.

We must certainly start with the theological concept of grace, for it appears that the structure of the religious notion is able to account for the elaboration of the popular senses, though whether popular theology is derived from the reasoning of the learned or whether the latter have based their doctrines on the popular premises which they absorbed in infancy – an argument that has raged for a century with regard to other aspects of culture – is an issue that we need not raise now, but it is worth observing that, unlike honor, grace seems never to have quite forgotten its etymological roots. Without entering into the subtleties of the numerous varieties of grace and the excessively optimistic or pessimistic doctrines which have been condemned as heresy from time to time, we can give, as its starting point, the pure gratuitous gift of God.

Louis Ott (1955) defines it as "a free gift of God unmerited by men" ("un don gratuit de la part de Dieu et immérité de la part de l'homme"). According to St. Thomas Aquinas it is especially the gift of the Holy Spirit, but it is also, in the Pauline view, associated particularly with Christ whose death redeemed us from original sin. As pardon obtained through Him, it is the key to salvation.

It develops within us as an *habitus*,[4] an acquired disposition to cooperate with the will of God, and this involves human will also, upon which the will of God operates, in St. Augustine's opinion. It is a supernatural accident created in our nature in such a manner as to adapt it to divine life, in the Thomist doctrine. Ott, again, says (1955: 314): "The unfathomable mystery of the doctrine of grace is to be found in the intimate collaboration and reciprocal intervention of divine power and human freedom. All the controversies and heresies regarding grace have their point of departure in this." ("Dans la collaboration intime et l'intervention réciproque de la puissance divine et de la liberté humaine se trouve le mystère insondable de la doctrine de la Grâce. Toutes les controverses et hérésies relatives à la grâce ont là leur point de départ.") And (pp. 348ff.) he discusses the relations between grace and freedom.

It is clear that the discussions center upon the role of human will which is insufficient by itself to attain salvation, but which cannot be dispensed with – without falling into determinism. If individual will is the essence of honor, the essence of grace is the will of God, which necessarily restricts the individual's will in some degree, but the attainment of grace can only be achieved with the cooperation of human will since God requires his benificence to be returned; the rites of the Catholic Church require the appropriate intentions on the part of the beneficiary in order to be valid. The Catholic solution to the paradox of theodicy resides in this mysterious conjunction of the will of God and the will of man.

Grace is connected with the will in another way also, for it is associated with purity. To be "in a state of grace" is to be sinless, to be redeemed, through confession, from the state of sin into which our all too human will has led us. To be in a condition to receive grace one must be cleansed of sinfulness, not only of original sin, of which one was discharged by the grace of the Holy Spirit at baptism, but of the sins which result from a will inadequately accorded to the will of God. Yet such provisions are necessarily insufficient to assure the attainment of grace, since the will of God on which this depends is by definition arbitrary. On account of this unpredictability, grace comes in popular usage to mean good fortune, that which cannot be foreseen.

Benveniste (1969), etymologist rather than theologian, discusses grace in less mystical terms, easier to grasp for one who is not versed in theology: "The Latin words show that, in origin, the procedure consists in giving service for nothing, without reimbursement; this literally gracious or gratuitous service provokes in return the manifestation that is called 'acknowledgment'." ("Les mots latins montrent que le procès, à l'origine, consiste à rendre un service pour rien, sans contre-partie; ce service littéralement 'gracieux' provoque en retour la manifestation que nous appelons 'reconnaissance'.") And later:

> Everything that refers to economic notions is tied to much vaster representations which bring into play the totality of human relations with divinities.
>
> Over and above the normal circuit of exchanges, that which one gives in order to obtain a counterpart, there is a second circuit, that of bounty and acknowledgment which is given without any consideration of a return of that which is offered, as an act of thanks.
>
> (Tout ce qui se rapporte à des notions économiques est lié à des représentations beaucoup plus vastes qui mettent en jeu l'ensemble des relations humaines ou des relations avec des divinités. Au-dessus du circuit normal des échanges, de ce qu'on donne pour obtenir – il y a un deuxième circuit, celui du bienfait et de la reconnaissance, ce qui est donné sans esprit de retour, de ce qui est offert pour "remercier").

Thus he distinguishes clearly between the two "circuits" of reciprocity, that which is properly called "exchange," an interchange of interests, and that which is inspired by a generous impulse, good will, gratuity, which demands only a reciprocity of sentiment. The former, subject to contract, can be simultaneous or protracted over a specified time. The latter, an exchange of grace, is simultaneous only in the handshake or the act of embracing, and normally requires an initiatory gesture, followed by a response, as in hospitality, or the return of favors. On account of its gratuitous nature there is no need, as in contractual exchange, to determine in advance what the value of the return shall be, nor when it shall be made, since none is envisaged, even though it may be hopefully expected. Whether any is to come depends not upon mutual agreement but upon the wishes of the recipient. The word "boon" expresses the notion of gratuity better than any other in English, for it is both a prayer, a request for a favor and the favor granted, a blessing and an unpaid service, such as ploughing or shearing, for a neighbor.

Though grace is a free gift of God, unpredictable, arbitrary and mysterious, there are nonetheless means of obtaining it: first of all through the sacraments. This is the main function of the rites of the Church, indeed of any church. Sacrifice is always a tentative to embark upon an exchange of grace with the Deity. The offering invites a return-gift of grace, the friendship of God, as it has been called. The Eucharist, the commensality of the mass, confession, prayer, and penance, the usage of "gratiferous" substances: incense, corn, wine, oil, and salt and water too, are all employed in the enterprise of obtaining grace, whether for the salvation of the souls of the faithful or their material prosperity. But the passage of grace is never guaranteed, even by the state of grace, the purity of intentions or the correct administration of the rite, because grace is a mystery which remains in the free gift of God.

If God is the source of grace, this does not mean that humans cannot generate it, and dispense it to others. The dictionaries' lists of meanings provide an abundance of examples, which we can leave the reader to examine at his leisure. The central core remains always the notion of gratuity, on the social as on the theological plane, and the essential opposition is to that which is rational, predictable, calculated, legally or even morally obligatory, contractually binding, creating a right to reciprocity. Grace is a "free" gift, a favor, an expression of esteem, of the desire to please, a product of the arbitrary will, human or divine, an unaccountable love. Hence it is gratuitous in yet another sense: that of being not answerable to coherent reasoning, unjustifiable, as when an insult is said to be gratuitous, or when a payment is made, over and above that which is due.

Thus it ranges from the tip left on the table for the waitress to the "golden handshake" of the company director who has lost his company a fortune through his bad judgment and swollen head, obstinacy or plain idleness, but who cannot be permitted to depart unrewarded for his 'long and loyal service," lest he do the company yet more damage by revealing its secrets. The tip displays the principles involved; it is called "a small gratuity" in self-consciously pompous language and this corresponds to our definition, for it is a sum paid over and above what is owed, whose amount is 'left to the appreciation of the client' as it used to be put on French restaurant menus, until, in blatant contempt of the very principle of grace, it became a compulsory charge of 10, and later 15 per cent. But the principle of grace is not so easily dismissed; the attempt to rationalize the system of payment and do away with the gratuity was unsuccessful, for no sooner had the clients become used to being charged more, than the custom crept in to leave a small tip extra in order to maintain their status relationship with those who served them.

George M. Foster's essay, "The analysis of envy" (1972) attributes the tip to the need to assuage the envy which a server might be supposed to feel towards those whom he serves, rather as the envy of witches is sometimes bought off by a gesture of gratuity, and there is certainly some substance to this explanation, but as a freely chosen expression of goodwill, the tip does no more than underline the absence of a contractual obligation, the opposition between that which is owed and that which is given and hence the superior social status of the giver, sealed by this petty gesture of patronization. For, if the offer of favor between equals awaits a return of grace, between unequals if it cannot be returned it can also mean a claim to superior status, and can thus even be resented by a person who does not wish to admit this claim.

In somewhat the same way a payment made under duress bears the opposite meaning in terms of honor to a payment made at the instigation only of the man who pays and who thus, provided it be accepted, accretes honor by his bounty. The economic result is identical in the two cases, but the social significance is the opposite. the man who is forced to pay is humiliated because he is shown to be inferior or dominated, the bountiful man demonstrates his superiority, his dominance, even though he may have been manipulated into giving his bounty. In each case the superior is the man who apparently imposes his will. Equally a gift may honor the recipient as an expression of sympathy and admiration or it may put him down if it can be interpreted to imply that he is in need of help. An excess of hospitality can humiliate as much as an insufficiency. For honor looks always up, pity, the poor relation of grace, looks down, but it is not always easy to tell which way people are looking; the same gesture can mean either homage or contempt, and it can even imply a measure of both simultaneously.

The techniques of begging illustrate this ambiguity well: they range from those who, sacrificing all shame, attempt to inspire pity by a display of their misery to those who propose a very expensive seat at a charity dinner they are professionally employed to organize. Who is helping whom? Is the donor of alms helping the beggar to stay alive, or is the recipient, of priestly caste perhaps, offering the donor the opportunity to acquire grace by becoming associated, through this act of charity, with the divinity? The ambivalence, so often encountered in the analysis of honor, as of grace, derives precisely from the possibility of giving different interpretations of the moral value attaching to an economically identical transaction. To offer a lady flowers is to pay her homage, to honor her – the donor has sacrificed to her the cost of the flowers, but she is none the richer, save in prestige – but if he offers her diamonds, she may be justified in wondering what are his intentions and whether she should accept. What exactly is the favor he hopes to receive in return?

Grace, positive and negative, in Grazalema

The connotations of grace are as varied as those of honor, the more so in Spanish than in English, it appears; for this reason, and also because I have previously examined the concept of honor in Andalusia (1977), I shall deal with grace as it operates in that part of Spain where its logic, thanks to its richness, can be seen more clearly. The variations in the daily use of grace, as of honor, distinguish different national cultures, and to some extent regions and classes as well, while the theological variants are matters of doctrine rather than usage and depend upon individual thinkers, cults and heresies – and historical periods.

Let us see then how the word *gracia* is used in the everyday life of a small Andalusian town, Grazalema (Pitt-Rivers 1954), starting with its role in popular theology. It figures in the subjectivity of the townsmen and, above all, townswomen as a state of grace, forgiveness subsequent to confession, the achievement of atonement which makes a person eligible to receive divine grace through the sacraments, or in answer to prayers. It is the essence of redemption through baptism; the rite of baptism assures the entry of a new-born child into the religious, that is to say, the human community. It expunges the sin of Adam through the application of the water, the salt, and the chrism, the unguent vehicle of grace which figures also in the last rites. It establishes the tie of spiritual affinity between the god-child and the god-parents who give the child its name. A person's first name is sometimes called his *gracia*. The god-parents, *padrinos*, are also called *padres de gracia*, spiritual parents or parents of the Christian name, in opposition to natural parents whose duties towards their children stand in marked contrast to those of the god-parents. The tie of spiritual affinity still exists in the popular view, with all the obligations of respect and the usage of the third person in speech between the *compadres* (parents and god-parents in relation to each other), despite the recent abolition of this tie in the dogma of the Roman Catholic (but not the Orthodox) Church (Pitt-Rivers 1976).

The grace of God is obtained through other sacraments, and prayers are offered to God and to the saints, above all to the various manifestations of the Holy Virgin, for their intercession to obtain divine favor in the form of personal advantages, cures, or miracles. Promises (*promesas*) were made to reciprocate the favor, if the prayers were answered, by wearing a robe recalling that of the image of the saint who had obtained the miracle or by performing a penance such as bearing a wooden cross, going barefoot or on knees during the processions in the celebration of Holy Week. The relationship with the saints is similar to that in Sicily described by Maria Pia Di Bella in her contribution to this volume (chapter 8).

Grace, like honor, is not at all the same thing for both sexes. While masculine honor is a matter of precedence in the first place and the man of honor strives to establish his name in the forefront of his group, the honor of women is rather a matter of virtue and sexual purity. The distinction is clearly marked in Sicily under the titles of "Name" and "Blood." The first is active and positive, a matter of attaining or inhering status and prestige or, in the plebeian community, the respect due to an honorable member. The second is negative and passive, a matter of avoiding any action that might stain the reputation of the family. Male honor is something to be won, increased, and defended against a rival; female honor is something to be conserved and protected from the evil tongues of the envious.

Grace is rather the reverse of masculine honor which depends upon individual will and ambition and this aligns it with female honor. In the first place women have, as it were, a prior claim to grace, not merely on religious grounds (they are more active in religious practice than men), but in the attribution of it in most of its forms. Aesthetic grace is purely feminine: men are not expected to have grace of movement, though they may dance with grace, and professional dancers are commonly assumed to be effeminate.

The *curanderas* (curers), also called *sabias* (wise women), possess an individual grace which comes to them from the saint of their devotion and enables them to cure through the grace in their hand and the performance of ritual gestures (signs of the cross on various parts of the patient's body), anointment with blessed oil, and prayers, muttered rather than declaimed because addressed to the divine powers rather than to the patient or the audience. The whole rite bears a general resemblance to baptism or to coronation, but the divinities to whom the prayers are addressed show the rite to be anything but orthodox Christianity. They include, apart from Jesus Christ and the Virgin Mary (in the manifestation which is the particular devotion of the *sabia*), the holy salt, addressed as a divine personage ("Salt, Salt, they call you salt, but I call you Holy Salt . . . "), the Seven Lions, Venus and Astarte, and others of an equally heterodox character. Aided by the holy oil, the evil leaves the patient's body and enters the arm of the curer who endures the pain for the next twenty-four hours or so.

Were a *sabia* to accept a gift of money in recompense for the cure she had performed, she would immediately lose her grace, for it belongs to God and would be withdrawn if she were to use it to obtain material gain. In return for her grace she can accept only symbols of grace, blessed candles to burn in front of the Virgin or holy oil, for she is operating upon that second circuit in which only grace can be returned for grace.

Given the ambivalence of magical powers, a general phenomenon in anthropology, it is not surprising that the first people to be suspected of witchcraft were the *sabias*. Could they not use their grace for evil ends also? Nothing appeared more plausible, for just as cursing is the reverse of blessing, so magical damage is the reverse of magical healing. It suffices to invert the intention to produce the opposite result. Witches are commonly credited with the power to inflict damage on man or beast, to do love-magic, to foretell the future, to drive men mad, etc., in brief to operate upon the health or sentiments of their victims.

Witchcraft everywhere depends upon envy, the will to do harm. But to do harm to an enemy is quite legitimate conduct; it is part of the game to take vengeance against the man who has damaged you or your reputation in order to recuperate your honor, "restore it to a state of grace." Witchcraft, however, is the attempt to take vengeance, not openly, but by covert dishonorable means, negative grace, invoking the aid of the Devil. For this reason the curer who enjoyed the greatest popularity in Grazalema never tired of repeating that she did "nothing but the things of God" and gave every outward sign of her piety.

In brief, witchcraft can here, as anywhere else moreover, be defined as an *habitus* of negative grace. Just as positive grace is opposed to reasoned empirical means of doing good, so negative grace is opposed to overt vengeance and employs the supernatural. The grace of the *sabias*, the source of their magical power, was something inherent in their bodies, like their sex. And it must be added that the female sex produces maleficent manifestations at the time of the period. Other forms of involuntary evil attributed to certain women were the Evil Eye and that quality, called *calio* in Grazalema, which makes them permanently dangerous as though they were menstruating. (A similar belief is found elsewhere in the region, though under different names, which all however recall the notion of heat: Pitt-Rivers 1954, where a more complete account is given.)

Witches can in fact be of either sex, though in Grazalema their techniques are totally distinct. Women cure by their grace which comes from God and if they bewitch it is, implicitly, by a misuse of that grace which has power even when perverted to nefarious ends, rather as the black mass, by inverting the symbols, uses the powers of religion to accomplish the work of the Devil.

Men have no grace in this sense; they cure and they bewitch not by the mysterious power of grace but by techniques which are supposedly rational and depend upon knowledge (though this is not general throughout Spain; men who cure by grace are to be found in the northern half of the country and as far south as the provinces of Alicante and Caceres). Thus apart from the practitioners of "scientific medicine," doctors and chemists, all of them men, those of folk medicine were limited to bone-setters and herbalists (the latter could also be women). The techniques of witchcraft practiced by men consisted of spells, and above all of spells to invoke the aid of the Devil through reading a book of magic. Since poltergeists were the work of the Devil, they were necessarily caused by "reading" and consequently by men only, though their victims were all women. Thus the distinction established by Evans-Pritchard between witchcraft and sorcery with regard to the ethnography of the Azande was exemplified in Grazalema, with the additional rider that witches were all, and by definition, female, while sorcerers were all male. None of the *sabias* knew how to read. Newspapers and notices on the municipal noticeboard were the main reading matter, that is to say, reading concerned the male sphere of relations with the administration and news of the outside world. Hence reading was an essentially male activity. News regarding the inner sphere of the community was passed by word of mouth at the fountain whither every household had to send a member every day to wash clothing if they wished to keep up with events.

Other manifestations of grace are accorded to people on different grounds. The signs of grace are various: the seventh child of a seventh child or, according to Brenan (1957: 99), the ninth child of a woman who has only male children, those born on a Friday, especially Good Friday, etc. The signs of grace are not confined to women, but they tend to reinforce its connection with the female sex: to begin with, in popular theology the source of grace was the Virgin Mary, who is announced as such in the Ave Maria. Hence all women named Maria have grace and this accounts for half the female population, for they are named after a specific manifestation of the Virgin, and are generally called, for abbreviation, only by the name of that manifestation. Hence Dolores, Carmen, Luz, Pilar, Mercedes, Milagros, Imaculada, and the place-names of her appearance, Lourdes, Fatima, etc., are all in fact Maria. But in any case it is said that all women are Mary simply by virtue of their sex. Good luck is also grace, prosperity on earth as well as salvation in the hereafter, for they are all given by God. In this sense it corresponds to *baraka*, as in a number of others which the reader may have noted.

"To have grace" also refers to the possession of natural "gifts" for they too are given by God (in classical mythology they were given by the Three Graces). Grace in this sense is the power to operate upon others, pleasing or amusing them or compelling their admiration and assent, hence it means not only grace in movement, as in English, but charm, wit, humor, a knack for something which others may not succeed in acquiring. Thus it is located in parts of the body: in facial expression, in the tongue, in the head, or in the hand. As such it is worth more than mere beauty (though beauty also is grace in the sense of being an unaccountable favor of nature) because it is more active than beauty which is passive and therefore less compelling than charm and wit. This is made clear in an Andalusian *copla* (popular verse form), put into the mouth of a highly self-confident young woman (which I translate very freely):

> Mi novio tiene una novia
> que es mas bonita que yo
> Mas bonita, si, serà
> Pero mas graciosa, no!
>
> My boy-friend has a girl-friend
> Who is prettier than me
> Prettier, yes, she may be
> But nothing like as much Fun
> [literally: but more *graciosa*, no!]

The greater value of grace as opposed to beauty is a commonplace of the folklore, underlined in rhymes such as: "La suerte de la fea, la guapa la desea" (The luck of the plain girl, the pretty girl envies her). As charm, the power to please, it is as mysterious as religious grace, and like this it belongs to the above, as a gift of God. To lack grace in this sense is to be *pesado*, literally heavy, hence boring. As in the work of Simone Weil, weight is the antithesis of grace – which contravenes all laws including the law of gravity.

As an unpredictable boon, "the grace of God" is used to refer to sunshine in the north of Spain and to rain in the south.

It is from this general sense of grace as the power to operate upon the will of others that Max Weber derived his conception of charisma, an extraordinary power of personal leadership. He treated it "as a property attributed to great innovating personalities who disrupt traditionally and rationally-legally legitimated systems of authority and who establish, or aspire to establish, a system of authority claiming to be legitimated by the direct experience of divine grace." (Shils 1968: 386).

It should be noted, however, that Weber did not conceive of charisma, as popular usage does today, uniquely as a type of political power, but as a much more general principle of social and cultural organization related to individual freedom and innovation. A "distinctive moral fervor" lies at the root of the attachment to charismatic authority which contrasts with submissive attitudes to traditional authority and sober calculation with regard to it. The general opposition between the ordinary and rational, on the one hand, and charisma, which is the essence of grace, on the other, assimilates the latter to the sacred, but at the same time the rejection of formalized forms may inspire "a fear and hence an opposition to the sacred itself" (Eisenstadt 1968: xviii).

But Weber recognized that charismatic and traditional authority come together, within the framework of social life, in "the routinization of charisma." In the same way the routinization imposed by the rules governing the rituals of the Church do not suffice to negate the charismatic power of their sacred origins. Hence the grace, inherent in the royal rituals described by Lafages, endows a very traditional form of government with the moral fervor associated with innovative forms, but, thanks to its fundamental ambiguity, it can equally well be acquired by rebels and used against sacralized royal authority, for, as Weber stressed, charisma contains destructive as well as creative tendencies. Here we have the solution to the paradox that, despite the superlative moral value of the monarch's coronation at Reims, rebellion against the legitimized royal ruler was always possible from time to time, for no ritual can guarantee the passage of grace. This ambiguity, admirably discussed by Eisenstadt, which its opposition to rationality engenders in charisma, is all the more striking in the instance of the *baraka* acquired by the successor to the Sultanate of Morocco in Jamous' essay: it is his *de facto* success in defeating his rivals in war that provides proof of the new Sultan's *baraka*, yet his charisma becomes routinized once he has succeeded to the throne. In such circumstances "the dichotomy between the charismatic and the orderly routine of social organisation seems to be obliterated" (Eisenstadt 1968: xxi).

The routinization, not to say the abuse, of grace reaches its peak in the formulae of social etiquette. In Spain "Your Mercy" (*Su Merced*) was used formerly as a form of address to a person of high status, just as in English "Your

Grace" is the title of a Duke or Archbishop (Pitt-Rivers 1974). Those who come immediately after the sovereign, "His Gracious Majesty," the source of all earthly grace, as of all honor (in whose gift is pardon, for he, alone, is above the law) participate, through their proximity to him, in his gracious nature. The title of Your Grace was also used formerly to royalty itself and the association was still extant in Shakespeare's day: "to have his daughter come into grace," i.e., marry into the Royal Family (*A Winter's Tale* IV.iii).

To summarize, in the sense of benefaction, gift, demonstration of benevolence, concession, graciousness, pardon, or indulgence, grace is inspired by the notion of something over and above what is due, economically, legally, or morally; it is neither foreseeable, predictable by reasoning, nor subject to guarantee. It stands outside the system of reciprocal services. It cannot be owed or won, specified in advance or merited. Hence it can mean remission of a sin or a debt, mercy, pardon, or forgiveness and thus it is opposed to justice and the law. As gratitude it is the only return-gift that conserves the nature of the initial prestation. You cannot pay for a favor in any way or it ceases to be one, you can only thank, though on a later occasion you can demonstrate gratitude by making an equally "free" gift in return. Like hospitality, which is a manifestation of it, or like violence, its contrary, a demonstration of malevolence, it can only be exchanged against its own kind (Jamous 1981). To attempt to reply to violence by invoking the sanctions of the law is behavior not approved by the code of honor (Pitt-Rivers 1977).

Grace, then, allows of no payment, no explanation, and requires no justification. It is not just illogical, but opposed to logic, a counter-principle, unpredictable as the hand of God, "an unfathomable mystery" which stretches far beyond the confines of theology. The opposition is clear and applies in every case: grace is opposed to calculation, as chance is to the control of destiny, as the free gift is to the contract, as the heart is to the head, as the total commitment is to the limited responsibility, as thanks are to the stipulated counterpart, as the notion of community is to that of alterity, as *Gemeinschaft* is to *Gesellschaft*, as kinship amity is to political alliance, as the open cheque is to the audited account.

Of course accounts are kept in fact much more often than is admitted, even if they are no more precise than a vague tally of favors, made in recollection. A lady of the bourgeoisie of south-western France remarked to intimates that she was terribly behindhand with her invitations; she "owed" God knows how many dinners! But the peasants of the same region never admit to keeping an account of the favors they have done in lending a helping hand to their neighbors, and one cannot help wondering whether it was not once so in England also, given the expressions "boon-ploughing" and "boon-shearing."

Gratuity as insurance in rural south-western France

It is well known that traditional small farmers in Europe practice a system of reciprocal service in agriculture and in general within the social life of the community. In fact there is great variety in the form this takes. The first anthropological account of such a system, in western Ireland, was given by Arensberg and Kimball (1949). "Cooring" was a relationship in which "friends" exchanged help on their farms. But the authors discovered that to be classed as a "friend" one had to have a tie of kinship. Those who had no kinsman in the community had no friend with whom to cooperate. A more recent account of another such system was provided by Sandra Ott who studied a French Basque sheep-farming village where the ties of neighborship are defined in terms not of kinship but of space. The roles of "first neighbor" and of the other "neighbors" depend upon the geographical position of their farms.

In "Magnac," the name I gave to a village in the Quercy in south-western France, neighborly cooperation depends upon a system of undifferentiated exchange, based upon the moral principle that neighbors must help each other. When asked to lend a hand one must, if at all possible, agree to do so, and under certain conditions help may even be proposed unrequested. This helping hand is lent without payment, without any reciprocity envisaged, and if the helper is asked what he is owed, he replies "nothing at all"; he explains that he has come to lend a hand in a quite disinterested spirit, because "we must all help each other," and he adds by way of explanation the traditional phrase, "Je ne suis pas regardant," which might be translated as "I don't keep accounts". It is the denial that any obligation has been incurred by the recipient of the service.

But small farmers have very good memories and the less they read and write the more they depend upon them. However, to admit to keeping accounts obliterates the virtue of this good-hearted gesture, just as recognizing that the intention of a gift was to provoke a counter-gift destroys the grace conveyed by it. Accounts are for public reckoning and contractual relations only, not for the exchange of favors. Where reciprocity is governed by specific rules defining the duties of each party the equivalence of the contribution of each can easily be established, but where there are no duties, no time limit, and the partners are unspecified, they cannot. The moral imbalance remains until the helper makes a request that will balance the unmentionable debt and there is no way he can be pressured to do so. Thus prescribed exchange is opposed to unspecified and open-ended reciprocity, just as market values and immediate reciprocity are opposed to the exchange of grace. Here we have a system of exchange of favor manifested through mutual services, rather than an exchange of services, that is to say, a demonstration of community solidarity which involves the sacrifice of immediate individual interests in favor of long-term collective ones (Bloch 1974).

The technological conditions for such a system have been explained elsewhere (Pitt-Rivers 1981). The harvest was traditionally the occasion for the display of neighborly relations when a considerable labor force was required to service the threshing machine. The day ended with a sumptuous meal, but the introduction of the combine harvester (which requires no work-force at all and works on contract paid by the hour) put an end to this institution. Heavy capitalization has unbalanced the system of reciprocity. The one equivalence that was implicit, if not stated, was that one man's labor was as good as another's. Even if this was not always true, the difference could not become apparent, since men worked in a chain forking the sheaves, which necessarily moved at the speed of the slowest man, but today, though the loan of machinery between neighbors has perhaps increased, the day of a man with a sixty-horse-power tractor is no longer the same thing as the day of a man with a horse. Nevertheless the system continues to function; the solidarity of the farming community struggles to maintain itself, the ideal remains intact.

Great prestige attaches to a good reputation as a neighbor. Everyone would like to be in credit with everybody and those who show reluctance to lend a hand when they are asked to do so soon acquire a bad reputation which is commented on by innuendo. Those who fail to return the favor done to them come to be excluded from the system altogether. Those of good repute can be sure of compliance on all sides. Thus the imbalance of favors provides the basis of an incipient system of patronage. The general opinion that so-and-so has "done much" for the community is the surest ticket for election or re-election in local politics.

But there is another factor which in a sense explains all the rest. A single family farm is particularly vulnerable to situations of crisis. If the husband breaks a leg who is going to milk the cows now that women no longer learn to milk? If the wife goes into hospital who is going to look after the children? The open-endedness of the system is justified ultimately as a mode of insurance. A good name as a neighbor is the most precious form of guarantee against disaster. In times of urgent need the neighbors come to offer their services. Who knows when they may not be in need of help themselves?

Grace in Kula: paradox in friendship and the gift

Community solidarity in the face of common dangers and difficulties prohibits the calculation of mine and thine. Yet the gift of one's time and energy is no different in moral terms from any other gift and it is therefore also subject to the paradox of the gift which descends from the theological mystery of grace – is not the doctrine of mystery theology's answer to an insoluble paradox? It is the same paradox as that of friendship. It derives from the fact that a gift is not a gift unless it is a free gift, i.e., involving no obligation on the part of the receiver, and yet, as Mauss so well explained, it nevertheless requires to be returned. In French a gift is "offered" not "given," just as an offering to the gods is. It is an expression of esteem, an honoring gesture, and as such complete in itself, requiring only to be accepted in order to achieve its end, which is, as in sacrifice, to convert "having" into "being," the dispossession of self in favor of another; the article or service which the donor sacrifices is transformed into becoming associated with the recipient, establishing or maintaining a particular tie with him.

It cannot be made with the intention of provoking a return-gift, for then it is not a free gift. If put to calculated ends, a gift is false charity, a manoeuvre abusing the notion of generosity. There is then no dispossession of self; the grace does not pass. But on the other hand, if it is not made in the hope of a return of the favor shown (which can itself only be shown by a reciprocatory gesture), then it is not a true gift either, but an act of largesse, a gesture of pity or a demonstration of superiority or power over the recipient. This was what Timon of Athens failed to understand. He made the mistake of humiliating his friends by an excess of generosity and he got what he deserved. Like the code of honor in which it might be taken to figure as a clause, the gift is shot through with ambivalence; any gratuitous gesture can "mean" one thing or its contrary according to circumstances. And the interpretation remains always debatable.

The gift, unrequited, implies either that it has been accepted as the receiver's due, a repayment or a tribute to him that he deserved and which consequently requires no recompense, but only amiable thanks, or that the recipient is unable to return it because he is not up to the standard of the donor and unwilling to appear inferior through making a return of lesser value. Finally, the gift, if not actually refused, may not be recognized at all. In the first case the giver is slighted because his gift has been taken for granted; in the second case, as in the potlatch, the receiver is humiliated because he has failed to "keep his end up"; in the third case the giver is rejected, for his favor is not welcome – the receiver does not wish to correspond, to initiate or to maintain his association with the giver. Possibly the latter was suspected of false intentions or pretentiousness in making such a gift in the first place.

If a gift is not returned the donor cannot complain, for he has placed his honor in the hands of the recipient. Were he to do so he would dishonor himself still further, for he would thereby imply that he made his gift only in order to obtain a hold on the receiver, to exploit him, to oblige him to reciprocate and in that case he deserved to be slighted.

The question of how soon the return gift should be made is also open to differing interpretations. That which is made too promptly stinks of payment, that is to say, it indicates a refusal to accept the favor as a free gift, a reluctance to remain in moral debt to a person whom you mistrust. (This over-hasty reciprocity is commonly found among persons who fear patronization by their social superiors.) It may equally, if returned immediately, mean the opposite: that the receiver is so grateful that he cannot wait to express his gratitude. The longer the delay the greater the confidence shown. On the other hand, to delay the return of a favor for too long may be taken to indicate, not the desire to show confidence in the sentiments of the donor and remain in his due until a worthy occasion presents itself to reciprocate, but indifference. Just as vengeance is all the more effective for being delayed – it is a dish that is to be served cold, so it is said – so gratitude is the more touching for having awaited the proper occasion to be expressed. But if the delay is too great the matter may appear to have been forgotten. In the meantime the receiver awaits his "revenge," as jokers sometimes say, playing on the ambivalence of the notion that the recipient can also be represented as "one down" in the competition to do favor.

In every case the interpretation depends upon the relative status of the participants, in terms of honor, power, prestige, morality, the nature of their relationship, and the customs of their milieu. In attempting to explain the paradox on which the notion of the gift is founded I have several times referred to Marcel Mauss and the reader may have noticed a certain indebtedness to his ideas at different points. However, note should be taken of the differences between our two approaches. Mauss was aware of no paradox. His aim was, briefly, to explain that, though gifts are "in theory voluntary, in fact they are given and repaid under obligation," "prestations which are in theory voluntary, disinterested and spontaneous, are in fact obligatory and interested. The form [of such exchanges] is that of the gift generously offered; but the accompanying behavior is formal pretence and social deception . . . " (1966). He demonstrated this by reference to extra-European and historical materials, whereas I have dealt mainly, if not quite entirely, with Europe. He makes no mention of grace, and the gratuity of the gift is, for him, a sociological delusion. Thanks have no legal or material reality and they do not come into the question. I believe that the gratuity of the gift is no delusion; however much it may have been abused, it is still the essence not only of sacrifice but of everyday gestures of favor.

The concept of grace which lay at the basis of my explanation cannot be assumed to exist elsewhere, even if Mauss' essay suffices to establish that the gift corresponds to a general sociological category. The basic opposition between human attempts to foresee and control the future through calculation and a law of contract, on the one hand, and the inscrutability of the will of God, whose grace is a mystery, on the other, is missing in primitive cultures. The Christian concept of grace is above all concerned with individual salvation, from the religious point of view, and this I put forward as the root of the extended senses of the word. It might be thought to follow therefore that there is nothing like grace to be found where there is no salvation, that is to say, outside Christianity. However we have suggested that, lacking such a fertile theological soil to grow in, a concept very similar to grace is found in Judaism (*hesed*) and in Islam (*baraka*). Reciprocity of favor, like friendship,[5] which is nothing more than this, is universal whether between man and God or between men. It is the other term of the opposition that is missing; legal contract is specific to societies that keep written records. The Indians of the Pacific coast of Canada have barely more of a juridical system than the Polynesians and Melanesians with whom Mauss was equally concerned, yet he seems to speak of both as though they had: he credits the Maori with a juridical mentality and he underlines the obligatory nature of the duty to return a gift, using the word "obligatory" in its pristine legal sense (binding by oath, promise or contract) and maintaining that this is defended by sanctions. (The sanctions that he mentions are uniquely warfare and witchcraft.)

Mauss was interested in the evolution of the notion of contract whose history he had traced back to a religious origin in the oath. I cannot help suspecting that his interest in contract and his lack of a notion of grace have produced a slightly warped vision of the gift. The only sanctions that ensure the necessity to return a gift are magical, and social only in so far as the refusal to return grace entails automatically negative grace and the breakdown of sociation which in primitive society leads easily, if not necessarily, to hostilities. The same is true in our own world. The slight represented by the failure to return grace provokes sentiments of resentment (malevolence) which far surpass in their virulence those which the simple evasion of contractual obligations inspires. One supports with equanimity the attempt to swindle one, especially when it fails, whatever one may think of the swindler, but one's *amour propre* is wounded by the refusal to respond to the offer of grace, for much more than material interests are at stake: one's intimate self is damaged.

Mauss wrote no ethnography himself, but he was a wonderfully perceptive reader of the field reports of others. However, his readings of those which support his essay on the gift appear to be far from unquestionable. They scarcely confirm the conclusions he drew. Whether or not to reply to the challenge represented by the potlatch is surely a voluntary decision, a matter of capacities and will, rather than an obligation. It is no more obligatory than to pick up the glove of any one who chooses to challenge you. If you refuse the challenge you may, depending upon the circumstances, lose prestige, that's all. But you do not risk losing your life. The exploitation of the principle of reciprocity in order "to fight with property," as the Kwakiutl put it, is a style of rivalry which has been recognized in our own society. In the Kwakiutl case the grace, proffered in the hospitality of the feast, conceals a negative aspect of hostility that is always lurking beneath the surface of hospitable relations. Mauss attributed this to the obligation to return the gift, and I gave a slightly different explanation in "The Law of Hospitality" (Pitt-Rivers 1977, chapter 5). In fact, as we have seen, grace is always in danger of being changed into negative grace when the sentiments in which it abides are lacking.

The obligation to return a gift is set forth above all in the example on which Mauss founded his theory. The *hau*, "the spirit of the gift" among the Maori, demands to be returned to its initial donor, and the man who scorns this demand risks supernatural sickness or even death.

leaving aside the criticisms that have been leveled at the interpretations of *hau* by Mauss and by others – Sahlins (1974) examines them – I note that many of the senses of this multi-faceted concept (though by no means all) are analogous with either grace, *baraka* or *indarra*, all of which are highly varied, as we have seen. One finds in them the essence of the gift, its non-contractual nature and something like the soul of the donor contained in it. *Hau*, like grace and like *baraka*, is that which is in excess, left over or supplementary, transcending exact reciprocity. Like *indarra*, especially, it is good fortune, fruitfulness, the power of increase, the power of Nature that makes the plants grow. And like *indarra*, it is also the wind. Moreover it should be recalled that the demands of the *hau* are reinforced only by the sanctions of witchcraft, which it has been suggested is supernatural malevolence, negative or inverted grace. At the same time there are many senses of grace, etc., that are not given for *hau* which, like *mana*, another Maori concept (Pitt-Rivers 1974), can also reside in stones, a capacity that seems quite foreign to grace – until one considers the Stone of Scone which demanded to be returned from Westminster to Scotland a few years ago (or so some students of St. Andrews University seemed to think).

I am not pretending that *hau* can be translated as grace, *baraka*, or *indarra*, but only that it is a concept of the same order, with all the differences that are implied by the contrast between a monotheistic and a polytheistic mode of thought. But it is no good trying to explain *hau* by an analogy with our own economic (or juridical) institutions when it is clearly closer to the religious ones that Benveniste found to be their source. One should rather take the totality of usages and search for their common qualities. Just as I have pointed to the similarity between honor and *mana*, so I can see that *hau* can be, at least in many contexts, assimilated to grace. It stands as a guardian of the second circuit, the embodiment of the abstract principle of gratuity.

I shall take for more detailed scrutiny the Kula gift-exchange of the Trobriand Islanders, not only because Mauss gives it as one of his main examples, but because it is, perhaps thanks to Malinowski, the most widely known ethnographic phenomenon in anthropological literature. In this elaborate system gifts of individual, named arm-shells and necklaces, which have no practical utility and in fact are seldom worn, circulate in opposite directions round a ring of islands over a span of several years and serve as the armature of a number of other more utilitarian exchanges and indeed of the whole system of social relations between islands.

Unlike the Nuer who, as Evans-Pritchard explained (1956: 223–4), were always cheated by Arab traders because they had no understanding of money and commercial transactions and thought they were establishing personal relations through an exchange of gifts, as in sacrifice – "a relation between persons rather than between things. It is the merchant who is bought, rather than the goods" – the Trobrianders understood trading very well, were skilled in barter and bargaining, though they had no currency, but did not confuse the counter-gift that is *owed* in payment with the ceremonial exchange of Kula. "The Kula exchange has always to be a gift, followed by a counter-gift; it can never be a 'barter', a direct exchange with assessment of equivalents and with haggling" (despite Mauss 1966: 187, where it is classified as "ceremonial barter"). "There must always be in the Kula two transactions, distinct in name, in nature and in time" (ibid.: 352). "There are means of soliciting, but no pressure can be employed." The valuable given in return is "under pressure of a certain obligation" but the only sanction is the partner's anger (ibid.: 353). Kula is "sharply distinguished from barter, gimwali." The equivalence of the counter-gift is never discussed, bargained about or computed (there is no standard by which this could be done anyway, ibid.: 359); it is "left to the giver and it cannot be enforced by any kind of coercion"; there is no direct means of redress, "or of putting an end to the whole transaction" (ibid.: 95, 96) nor indeed to the specific Kula-partnership. What kind of a contract is this? "Kula is never used as a medium of exchange or as measure of value," "the equivalence of the values exchanged is essential, but it must be the result of the repayer's own sense of what is due to custom and to his own dignity" (ibid.: 511).

Kula objects are offered to the spirits (including land-crabs) to make them benevolent, "to make their minds good" (ibid.: 512). The choice of which valuable to offer to a given partner is free, but the owner can be influenced by solicitory gifts of food or petty valuables, by "rites and spells, acting on the mind of a partner to make him soft, unsteady in mind, eager to give kula" (ibid.: 102). Resentment caused by disappointment at not receiving a given famous arm-shell or necklace cannot be expressed openly and directly, but it can be avenged by black magic; "many men died because of [such-and-such a necklace]" (ibid.: 359), i.e. by witchcraft.

When unable to find a suitable return-gift to offer to a partner one may make an "intermediate gift" in order to avoid the accusation of being "slow" or "hard" in Kula, but this is "a token of good faith," not, as Mauss thought, interest on an outstanding debt. The definitive proof of this, if needed, is that an equivalent return is expected for the intermediate gift (ibid.: 98). In a word, the Trobriand Islanders distinguish as clearly as Benveniste between the two circuits and in Kula there is no trace of contract, nor of sanctions other than the spoiled grace which is sorcery and the desire for vengeance which is allowed no overt expression. The bad heart of the unsatisfactory partner can be dealt with only by magical charms. The only obligation on the circuit of grace, a moral one, is the return of grace. Kula valuables are vehicles of grace like the chrism of baptism or the oil of the Holy Ampulla; they have no material utility and no value in relation to utilities. A Kula partnership is no more contractual than the relation between blood-brothers in Africa or *compadres* in Spain or Latin America and like them it contains no possibility of annulment. A contract, on the other hand, to be valid in English law must specify its duration.

Hence one might conclude by inverting the proposition of Mauss, saying that the gift is not in theory voluntary while in fact obligatory, but in theory – according to the theory of Mauss – obligatory while in reality voluntary, for it depends uniquely upon the will of the partner; nothing is specified with regard to the return gift, what it should be or when it should be returned. The moral obligation is only to return grace and what is resented if it is not returned is not the material loss but the rejection of the donor's self. It is always the thought – look back to its etymology – that counts.

If the anthropological study of friendship was so slow to be broached it was perhaps not unconnected with the failure to recognize the importance of grace. For grace is the essence of friendship. It is defined by the theologians as "the friendship of God" and on the terrestrial plane it remains the essence of that institution, as its subsidiary vocabulary makes clear: bounty, boon, favor, gratuity, thanks, etc. Since reciprocal favor is the basis of friendship and the exchange of gifts its mark, it is no surprise that the paradox of the gift, as I have explained it, is present in friendship which, like the gift, and like honor also, possesses both a subjective aspect, that which is felt in the heart and which determines the will, the aspect that relates it to the sacred, and the objective aspect which relates it to material interests, to others and to society in general. The paradox is that, while friendship to be true should be inspired by sentiment only and disinterested, yet it must be reciprocal not only in sentiment but in material gifts and services, for sentiments must be expressed by deeds. Hence it must be disinterested in motive, but profitable all the same, since the test of a friend is that he devotes himself to your interests. Actions speak louder than words.

It is vain to attempt to split the paradox and found a typology of friendship upon its two terms, distinguishing between the "affective" and the "instrumental" types, as Eric Wolf proposed in a well-known essay, for how can the ethnographer know whose feelings are pure and whose are not? It is hard to guarantee the former in the face of the constraints of daily life, but no one admits to the latter. The distinction then rests upon a factor that is exceedingly difficult to establish. The same difficulty faces the Church in determining the validity of its rites: how to know the true intentions of the participants. But the Church at least, through the Court of the Rota, summons evidence and goes through a judicial process before giving judgment, while each man must judge the heart of his friend for himself.

Aristotle in the Nicomachean Ethic did better than anyone since at disentangling the two aspects, by distinguishing between friends who are useful, friends who are amusing, and friends who are loved, but on account of the paradox it is not normally made clear who is to be classed as which and it is always a disappointment to those who thought they were loved to discover that they were merely useful. Crossed lines between the two circuits always lead to trouble (Pitt-Rivers 1984). Once more, the uncertainty of grace, one of its essential aspects in the religious sphere (and one which gives rise to the sin of presuming on the grace of God), is demonstrated in everyday social relations.

Honor and grace: their mediative function

We have tried to encompass the totality of grace and explain the coherence of the whole concept and the relation between its different meanings. In defining a notion of such variety, it is not enough to study only the concept by itself, one must place it within sets of wider range and look for its connections with others. The one of particular interest for this book is that which forms, with grace, its title, honor, for it seems to deal with problems in the same field: the destiny of a man and his relations with other people and with God. It shares with honor the same tendency to be evanescent and self-contradictory. Sometimes the two words are almost interchangeable: you pay honor in offering grace, for it is an expression of sentiment freely willed. You expose your honor in doing so, precisely because there is no obligation to return grace unless it comes from the heart – and you are dishonored if you get a "brush-off," in which case you are justified in being offended. Hence it can be seen that exchanges of honor are very similar to exchanges of grace. To be favored or privileged is to be honored, exalted above those who are not favored. Rituals of honor are continually marking distinctions of this sort.

Grace, as a verb, underscores the similarity (O.E.D.): "To confer honour or dignity upon, do honour or credit."

Grace, like honor, is associated with power and with royalty, as we have seen. To lose power is "to fall from grace," "to be put down," "to go out of favor," "to be disgraced."

Moreover, physical contact is all important in the conveyance of both: one is dubbed a knight by a touch on the shoulder with the sovereign's sword; one ingests the host; one is "touched" by grace; the transmission of grace at baptism involves the application of blessed substances to the body of the initiate. Anointing, whether of kings or neophytes, is a ritual technique for according grace. The affront to honor is expressed by a slap on the face. The rituals of dishonor involve the physical person also: the stocks, the stripping of honorific symbols, the defilement of the head, cutting the hair, marking the body in castigation; honorific status is inscribed upon the body. All these rites relate to the person-in-himself where the will is centered.

In other contexts grace appears not similar to but complementary to honor, as in the coronation ritual analyzed by Lafages in chapter 2: the king is inducted simultaneously as knight and bishop, temporal and spiritual leader, fount of honor, whose legitimacy is guaranteed by grace.

They can also be seen as alternative bases for renown as in Di Bella's account of the relation between Name and Blood: titles to honor, on the one hand, and miracles, manifestations of divine grace, on the other. Miracles step outside the worldly hierarchy and by appealing to the principle of anti-honor achieve more than can be won in the competition for precedence.

Finally there are those contexts in which honor is frankly opposed to grace, as for example in the sense that honor depends only upon the individual will to win, while grace depends upon the will of God and cannot be won.

Grace (O.E.D.) is "hap, luck or fortune," while honor can be saved when all else is lost by fortitude and determination; each man is guardian of his own honor, for it depends upon his will, his worth as a person, or his rank. Grace ignores moral and social qualities. It is arbitrary as the divine whim. Honor implies combat, rivalry, and triumph, while grace means peace.

The unverifiability of intentions and the state of the heart, which are necessary to both honor and grace, the paradoxes which assail both and the uncertainty of divine judgment which refuses to submit to mundane reasoning, have encouraged societies with writing to replace the reciprocity of the heart by the law of contract and provide sanctions for its enforcement. Taking reciprocity out of the field of grace detaches it from the sentiments and objectifies it, making it abstract and depersonalized. As soon as the monarchy was strong enough, it suppressed the judicial combat and took justice out of the unpredictable hands of fate, curtailing the problem of theodicy. Yet the affective side of life cannot be obliterated. Despite their contrary aspects grace and honor remain and contribute each to the composition of the other.

When the conquistador went to America "to win honor with my lance in my hand" and at the same time to get the gold needed to support such pretensions to honor, he depended upon his own ambition, courage, toughness, and wiles, but once he had risen to eminence, power, and wealth, he had the means of dispensing grace. Only those with power can favor or pardon; you must possess a thing before you can give it away; the beggar can return grace only in the form of a blessing and the hope that God may reciprocate the gesture of charity.

But grace is closer to God than the ambition to succeed, and for that reason he who has renounced all pretension to honor by demanding (and exploiting) pity enjoys a privileged relationship with the Almighty. Grace, not honor, is the ideal enjoined by the Beatitudes. Indeed the contradiction is spectacular, the lesson clear: one must renounce one's claim to honor as precedence if one is to attain the fellowship of the Holy Ghost, or more precisely one must invert it, adopt the counter-principle represented by the honor of women, whose sex excludes them in theory from the agonistic sphere. (The resemblance between feminine honor and grace has already been noted.) In somewhat similar vein the sacred status of priests imposes upon them the renunciation of violence and worldly ambition as well as the right to play the male role in relation to women.

One might be tempted to jump to the conclusion that this opposition between masculine honor and grace is a product of the moral revolution imposed by Christianity – this was what Nietsche thought – yet grace is much older than Christianity and Mauss (1966: 15) had already noted the generic connection between the gods and the poor through whom they can be approached; we also know from classical Greece that Zeus was accustomed to adopt the disguise of the beggar who enjoyed his special protection. The opposition is therefore more general than its connection with Christianity. The sacred status of beggars can be found in India also.

Moreover, this apparent anomaly is not without solution. To begin with, in the logic of agon, rivalry is wedded to equality, for one must feel oneself to be the equal of one's antagonist in order to compete, yet the test of strength aims always to destroy this equality and establish hierarchy: a victor and a vanquished (Pitt-Rivers 1960). The latter can demand revenge "when all is lost save honor," but, theoretically at least, they can no longer issue or accept a challenge if they are not conceived as equal. The institution of the handicap is the means of enabling them to do so, by making them equal.

However, once the hierarchy is established, the competition is over, and from that moment the man who has achieved power, preeminence, the dominant position, wishes only to legitimize it in the public consensus. His honor changes its nature and from striving to achieve precedence it becomes honor validated in the view of others, honor as virtue, recognized by his subordinates and dependants, whom he protects. This simplistic paradigm of power would have no interest if it did not correspond to a transition from agonistic honor to grace which we can detect in other circumstances. It would appear that there are in Western civilization two opposed – and ultimately complementary – registers: the first associated with honor, competition, triumph, the male sex, possession, and the profane world, and the other with peace, amity, grace, purity, renunciation, the female sex, dispossession in favor of others, and the sacred.

The ambitious man, when he has succeeded, won honor, and achieved dominance, requires that the qualities by which he has done so be confirmed and consecrated, that what he has achieved be conserved for the future. Thus he must then show qualities that are the contrary: generosity and moderation (which cannot now be taken for weakness), forbearance (his status entitles him to ignore what would once have constituted a slight), wisdom and, in order to represent the collectivity, a regard for the feelings and opinions of others. We have given examples: those old Cypriot *sophrons*, described by Peristiany, whose ideal is to end up with with nothing, having distributed all their possessions among their children, or the builder of the hospital described by Caro Baroja, or the Sultan of Morocco who, having fought and slaughtered to win the succession, becomes a sacred figure, a man of peace, able to settle the quarrels of others in accordance with the norms of Islam, for feud and vendetta can be ended only by an appeal to the sacred which such persons represent. The *shorfa*, peace-makers of the Rif, do not enter the struggle for precedence, for they belong to sacred lineages, the guardians of the shrine which is a sanctuary.

The man of honor must in a sense convert his honor into grace and thus render himself impregnable, because legitimate, be loved by those he has defeated, make peace where once he made war, trade his *de facto* power for right, his dominance for status, his business acumen for magnanimity. He must struggle to achieve honor, but the proof that he has achieved it is that he no longer needs to struggle.

Hence that which appears at first as an anomaly if not a contradiction becomes, once viewed in the perspective of the life cycle, a relation of complementarity and in the end concordance, for, since grace is the will of God, it is by His grace that the triumphant hero received his endowment in the first place, was predestined to win. If one ends up dishonored, it was for lack of grace in the first place. Honor stands before the event; his honorable qualities produce the victor. Grace stands behind it; the will of God is revealed in the outcome. Each is therefore a precondition of the other. If you lack grace you will not attain honor, in which case you will lack the means to be gracious.

However, there are those who do not have to achieve honor and convert it into grace, for they are born with both. Louis XIV became king at the age of six and he had so much grace that he thought himself entitled to elevate his bastards over the heads of those whose kinship with him was more distant, if more conventionally respectable. As Leroy Ladurie makes clear, the Little Duke was not of that opinion! Their difference was understandable, for the royal bastard is always of ambiguous status: he has the honor attaching to his recognized descent, he was conceived by personal will, not for reasons of state, a "love-child," and he is usually his father's favorite, but he bears the stigma of illegitimacy. (Though let us remember that grace is opposed to and above the law.) Therefore royal bastards are superior in a covert, mystical sense, while being at the same time inferior, socially, to their legitimate kinsmen.

There are those also whose honor and grace is consecrated by less obvious rituals than a coronation. The administrators of modern France, former students of the Grandes Ecoles, win their honor by triumph over rivals in the entrance examination – personal excellence has replaced genealogical descent or wealth as the criterion of the elect – and thereafter, their intelligence and competence certified by their status as *normaliens, politechniciens*, etc., they remain, whatever they do or say, as Bourdieu explains in chapter 4, with a kind of grace, the living embodiment of the will, not of God but of the French Republic.

Both honor and grace are mediative concepts; they interpret events in accordance with the prevailing values of society, putting the seal of legitimacy on to the established order. Together they constitute the frame of reference by which people and situations are to be judged. They are indistinguishably blended in the Basque concept of *indarra*, in which natural, cultural, and social forces come together; the life force of men, plants and the wine, the grace of the Holy Virgin, and the legitimacy of property rights in houses. Elements of both honor and grace are found in Polynesian *mana*, and other "legitimizing concepts" can be found in the ethnographies of other parts of the world.[6] They supply the point of junction between the ideal and the real world, the sacred and the profane, culture and society.

Notes

1 Hertz 1960: 72; Mauss 1968: 201, 206; Lévi-Strauss 1962: 298: "Le schème du sacrifice consiste en une opération irréversible (la destruction de la victime) afin de déclencher, sur un autre plan, une opération également irréversible (l'octroi de la grâce divine), dont la nécessité résulte de la mise en communication préalable de deux 'récipients' qui ne sont pas au même niveau." Gilsenan also uses the word when explaining what *baraka* is (1973: 33, 35, 76), as do many other Islamists since Westermarck (e.g., Gellner 1973: 40): "it can mean simply 'enough', but it also means plenitude, and above all blessedness manifested among other things in prosperity"; Rabinow 1975: 2: "the central symbol of vitality in Moroccan culture", "source of charismatic authority and cornerstone of legitimacy," "the manifestation of God's grace on earth," (ibid.: 25).

2 In fact, however, the Nuer word *muc* which he translates as "God's free gift in return for a sacrifice" (1956: 331) appears to correspond exactly to the sense of "divine grace" as Lévi-Strauss used it in the passage quoted in note 1 which was largely based on Evans-Pritchard's ethnography.

3 An anthropologist's wife was reproved by her friends in the village where he worked for saying "thank you" too often. Did she not realize, they asked, that you don't thank friends for petty services; for to do so destroys the *confianza* which exists between friends. You should take such minor services for granted to show *confianza* (William Kavanagh, personal communication).

4 The term has been borrowed by the philosophers and, finally, by the social sciences. For its utilization in the latter, see Bourdieu 1987: 23: "a system of acquired schemas functioning practically as categories of perception and assessment or as classificatory principles as well as the organising principles of action."

5 Any reader who might be tempted to follow those authors who have maintained that there is no friendship, only kinship, in primitive society is referred to Firth 1936.

6 For example, the concept of *mahano* (Beattie 1960) has much in common with

mana, as noted by Balandier (1967: 119–30), who compares it with other African concepts. See also the Bouriate concept of *xeseg*, translated into Russian as "grace" in Hamayon (1987: 606–10, 620–5).

References

Arensberg, Conrad and Kimball 1949. *Family and Community in Western Ireland*, Harvard

Balandier, Georges 1967. *Anthropologie politique*, Paris

Bateson, Gregory 1972. Style, grace and information in primitive art, in his *Steps to an Ecology of Mind*, New York, pp. 128–52

Beattie, J. M. 1960. On the Nyoro concept of mahano, *African Studies*

Benveniste, Emile 1969. *Le Vocabulaire des institutions indo-européennes*, Vol. I, *Economie, parenté, société*, Paris

Bloch, Maurice 1974. The long term and the short term: the economic and political significance of the morality of kinship, in J. Goody (ed.) *The Character of Kinship*, Cambridge, pp. 75–87

Bourdieu, Pierre 1987. *Choses dites*, Paris

Brenan, Gerald 1957. *South from Granada*, London

Catedra, María 1981. Death as a cultural process, Ph.D. Thesis, Penn University

Catedra, María 1988. *La Muerte y otros mundos*, Jucar

Eisenstadt, S. N. 1968. Preface to Weber 1968

Evans-Pritchard, E. E. 1937. *Witchcraft, Oracles and Magic among the Azande*, Oxford

Evans-Pritchard, E. E. 1956. *Nuer Religion*, Oxford

Firth, Raymond 1973. Bond-friendship in Tikopia, in Dudley Buxton (ed.), *Custom is King: Essays in Honour of R. R. Marett*, London, pp. 299–327

Firth, Raymond 1973. Bodily symbols of greeting and parting, in *Symbols Public and Private*, London

Fortes, Meyer 1969. *Kinship and the Social Order: Lewis Henry Morgan Lectures*, London

Foster, George 1972. The Analysis of Envy, *Current Anthropology* 13 (2), 165–86

Gellner, Ernest 1973. Concepts and society, in *Cause and Meaning in the Social Sciences*, London

Gilsenan, Michael 1973. *Saint and Sufi in Modern Egypt: An Essay in the Sociology of Religion*, Oxford

Gusdorf, Georges 1948. *L'Expérience humaine du sacrifice*, Paris

Hamayon, Roberte 1987. La chasse à l'âme, le chamanisme sibérien à partir de l'exemple bouriate, doctoral thesis, Université de Paris

Hertz, Robert 1960. Représentations collectives de la mort, in *Mélanges de sociologie religieuse*, translated in *Death and the Right Hand*, London

Jamous, Raymond 1981. *Honneur et baraka: les structures traditionnelles dans le Rif*, Cambridge and Paris

Lévi-Strauss, Claude 1962. *La Pensée sauvage* (translated as *The Savage Mind*, Chicago 1966)

Lewis, I. M. 1961. *A Pastoral Democracy*, Oxford

Malinowski, Bronislaw 1922. *Argonauts of the Western Pacific*, London

Mauss, Marcel 1966. *The Gift*, London

Mauss, Marcel 1968. *Oeuvres*, vol. I, *Les Fonctions sociales du sacré*, Paris

Ott, Louis 1955. *Précis de théologie dogmatique*, Paris

Ott, Sandra 1981. *The Circle of Mountains: A Basque Shepherding Community*, Oxford
Peristiany, J. G. 1966. Honor and Shame in a Cypriot Highland Village, in J. G. Peristiany (ed.) *Honor and Shame*, London and Chicago, pp. 171–90
Pitt-Rivers, Julian 1954. *The People of the Sierra*, London (2nd edition, Chicago 1971)
Pitt-Rivers, Julian 1960. The Egalitarian society, in *Actes du Colloque International d'Ethnologie*, vol. I, part II, Paris
Pitt-Rivers, Julian 1967. Contextual analysis or the locus of the model, *European Journal of Sociology*
Pitt-Rivers, Julian 1973. The kith and the kin, in J. Goody (ed.) *The Character of Kinship*, Cambridge, pp. 89–105
Pitt-Rivers, Julian 1974. *Mana: An Inaugural Lecture*, London
Pitt-Rivers, Julian 1976. Ritual kinship in the Mediterranean: Spain and the Balkans, in J. G. Peristiany (ed.) *Mediterranean Family Structures*, Cambridge
Pitt-Rivers, Julian 1977. *The Fate of Shechem or the Politics of Sex*, Cambridge
Pitt-Rivers, Julian 1981. Lending a hand, Oxford conference on conflict and cooperation (unpublished)
Pitt-Rivers, Julian 1984–5. La revanche du rituel dans l'Europe contemporaine, in *Annuaire de l'Ecole Pratique des hautes Etudes*, 5th section, vol. XVII, reprinted in *Les Temps modernes*, 1987
Pitt-Rivers, Julian, in press. Le paradoxe de l'amitié, *Actes du Colloque de Palerme 1984*
Rabinow, Paul 1975. *Symbolic Domination: Cultural Form and Historical Change in Morocco*, Chicago
Sahlins, Marshall 1974. The spirit of the gift, in *Stone Age Economics*, London, pp. 168–83
Shils, Edward 1968. Charisma, in *International Encyclopedia of the Social Sciences*, Vol. II, Chicago
Weber, Max 1968. *On Charisma and Institution Building*, with Preface by S. N. Eisenstadt, Chicago
Weil, Simone 1948. *La Pesanteur et la grâce*. Paris
Wolf, Eric R. 1900. Systems of patronage and friendship, in Michael Banton (ed.) *The Anthropology of Complex Societies*, pp. 1–2

Index

Cambridge Studies in Social Anthropology

Editors: JACK GOODY, STEPHEN GUDEMAN, MICHAEL HERZFELD, JONATHAN PARRY

1 The Political Organisation of Unaymwezi
R. G. ABRAHAMS
2 Buddhism and the Spirit Cults in North-East Thailand*
S. J. TAMBIAH
3 Kalahari Village Politics: An African Democracy
ADAM KUPER
4 The Rope of Moka: Big-Men and Ceremonial Exchange in Mount Hagen, New Guinea*
ANDREW STRATHERN
5 The Majangir; Ecology and Society of a Southwest Ethiopian People
JACK STAUDER
6 Buddhist Monk, Buddhist Layman: A Study of Urban Monastic Organisation in Central Thailand
JANE BUNNAG
7 Contexts of Kinship: An Essay in the Family Sociology of the Gonja of Northern Ghana
ESTHER N. GOODY
8 Marriage among a Matrilineal Elite: A Family Study of Ghanaian Civil Servants
CHRISTINE OPPONG
9 Elite Politics in Rural India: Political Stratification and Political Alliances in Western Maharashtra
ANTHONY T. CARTER
10 Women and Property in Morocco: Their Changing Relation to the Process of Social Stratification in the Middle Atlas
VANESSA MAHER
11 Rethinking Symbolism*
DAN SPERBER. Translated by Alice L. Morton
12 Resources and Population: A Study of the Gurungs of Nepal
ALAN MACFARLANE
13 Mediterranean Family Structures
Edited by J. G. PERISTIANY
14 Spirits of Protest: Spirit-Mediums and the Articulation of Consensus among the Zezuru of Southern Rhodesia (Zimbabwe)
PETER FRY

15 World Conqueror and World Renouncer: A Study of Buddhism and Polity in Thailand against a Historical Background*
S. J. TAMBIAH
16 Outline of a Theory of Practice*
PIERRE BOURDIEU. Translated by Richard Nice
17 Production and Reproduction: A Comparative Study of the Domestic Domain*
JACK GOODY
18 Perspectives in Marxist Anthropology*
MAURICE GODELIER. Translated by Robert Brain
19 The Fate of Shechem, or the Politics of Sex: Essays in the Anthropology of the Mediterranean
JULIAN PITT-RIVERS
20 People of the Zongo: The Transformation of Ethnic Identities in Ghana
ENID SCHILDKROUT
21 Casting out Anger: Religion among the Taita of Kenya
GRACE HARRIS
22 Rituals of the Kandyan State
H. L. SENEVIRATNE
23 Australian Kin Classification
HAROLD W. SCHEFFLER
24 The Palm and the Pleiades: Initiation and Cosmology in Northwest Amazonia*
STEPHEN HUGH-JONES
25 Nomads of Southern Siberia: The Pastoral Economies of Tuva
S. I. VAINSHTEIN. Translated by Michael Colenso
26 From the Milk River: Spatial and Temporal Processes in Northwest Amazonia*
CHRISTINE HUGH-JONES
27 Day of Shining Red: An Essay on Understanding Ritual*
GILBERT LEWIS
28 Hunters, Pastoralists and Ranchers: Reindeer Economies and their Transformations*
TIM INGOLD
29 The Wood-Carvers of Hong Kong: Craft Production in the World Capitalist Periphery
EUGENE COOPER
30 Minangkabau Social Formations: Indonesian Peasants and the World Economy
JOEL S. KAHN
31 Patrons and Partisans: A Study of Two Southern Italian Communes
CAROLINE WHITE
32 Muslim Society*
ERNEST GELLNER
33 Why Marry Her? Society and Symbolic Structures
LUC DE HEUSCH. Translated by Janet Lloyd
34 Chinese Ritual and Politics
EMILY MARTIN AHERN
35 Parenthood and Social Reproduction: Fostering and Occupational Roles in West Africa
ESTHER N. GOODY
36 Dravidian Kinship
THOMAS R. TRAUTMANN

37 The Anthropological Circle: Symbol Function, History*
MARC AUGE. Translated by Martin Thom
38 Rural Society in Southeast Asia
KATHLEEN GOUGH
39 The Fish-People: Linguistic Exogamy and Tukanoan Identity in Northwest Amazonia
JEAN E. JACKSON
40 Karl Marx Collective: Economy, Society and Religion in a Siberian Collective Farm*
CAROLINE HUMPHREY
41 Ecology and Exchange in the Andes
Edited by DAVID LEHMANN
42 Traders without Trade: Responses to Trade in two Dyula Communities
ROBERT LAUNAY
43 The Political Economy of West African Agriculture*
KEITH HART
44 Nomads and the Outside World
A. M. KHAZANOV. Translated by Julia Crookenden
45 Actions, Norms and Representations: Foundations of Anthropological Inquiry*
LADISLAV HOLY and MILAN STUCKLIK
46 Structural Models in Anthropology*
PER HAGE and FRANK HARARY
47 Servants of the Goddess: The Priests of a South Indian Temple
C. J. FULLER
48 Oedipus and Job in West African Religion*
MEYER FORTES
49 The Buddhist Saints of the Forest and the Cult of Amuylets: A Study in Charisma, Hagiography, Sectarianism, and Millennial Buddhism*
S. J. TAMBIAH
50 Kinship and Marriage: An Anthropological Perspective (available in paperback in the USA only)
ROBIN FOX
51 Individual and Society in Guiana: A Comparative Study of Amerindian Social Organization*
PETER RIVIERE
52 People and the State: An Anthropology of Planned Development*
A. F. ROBERTSON
53 Inequality among Brothers: Class and Kinship in South China
RUBIE S. WATSON
54 On Anthropological Knowledge*
DAN SPERBER
55 Tales of the Yanomami: Daily Life in the Venezuelan Forest*
JACQUES LIZOT. Translated by Ernest Simon
56 The Making of Great Men: Male Domination and Power among the New Guinea Baruya*
MAURICE GODELIER. Translated by Rupert Swyer
57 Age Class Systems: Social Institutions and Politics Based on Age*
BERNARDO BERNARDI. Translated by David I. Kertzer

*available in paperback